Fieldwork Projects
in Biology

FIELDWORK PROJECTS
in BIOLOGY

Marjorie Hingley, M.Sc.

BLANDFORD PRESS
Poole   Dorset

First published in the U.K. 1979

Copyright © 1979 Blandford Press Ltd,
Link House, West Street,
Poole, Dorset, BH15 ILL

ISBN 0 7137 0964 2 ✓

British Library Cataloguing in Publication Data
Hingley, Marjorie
   Fieldwork projects in Biology.
   1. Biology—Fieldwork
   I. Title
   574′.07′23   QH318.5

Filmset in $11/12\frac{1}{2}$pt Ehrhardt and printed by
BAS Printers Limited, Over Wallop, Hampshire

Contents

Acknowledgements

I am most grateful to past and present pupils who have taken part in and helped to shape many of these projects. My colleagues in the Biology Department at Cranbrook School have given me every encouragement for which I thank them warmly.

My thanks are also due to Dr B. F. Gill of Edge Hill College of Higher Education for permission to include his method for measuring dissolved oxygen using a modified Winkler's technique. This was originally published in the *School Science Review*, Vol. 58, No. 204, March 1977.

The table on page 77 has been reproduced from Hale *The Biology of Lichens* published by Edward Arnold 1972.

The Appendix diagrams were drawn by Christine Taylor.

Marjorie Hingley

Introduction

How to use the book

The suggestions in this book are intended to provide the basis for short and long-term fieldwork projects. Their main use is probably for the optional A and O level projects of the various examining Boards. It is strongly recommended that such projects are commenced at the beginning of the summer term in the year previous to sitting the examination. This is especially important where fresh water projects are chosen, or where a whole year's observations are required. Some are also suitable for Natural History Societies, Outdoor Pursuits Centres, Duke of Edinburgh Award, Scouts and Guides as well as younger age groups. They can be adapted and extended to suit most age groups and levels of ability. Many have been tried out over thirty years teaching experience.

Students often find it difficult to think of suitable ideas; they may choose ones which are too difficult, or lead nowhere. On the other hand they do not always appreciate those ideas suggested by the teacher or group leader. They may accept the idea at first, but when the going becomes hard they often want to change direction. This may waste a considerable amount of time and energy for everyone concerned. In some cases of course it is a necessary preliminary for settling down to something more congenial.

Teachers may find ideas in this book which they can use as they stand, or adapt to their particular needs, or to available habitats and material. Students may browse through the contents and hit upon some attractive idea. Most students need more individual help than a busy teacher can give. This book supplies some of the information and guidance required to get off to a good start.

The most important ingredient for the enjoyment and success of any fieldwork project is curiosity. This should be backed up by careful observation and accurate recording. A field notebook and a meticulous labelling procedure are recommended. There is nothing worse than relying on the memory or messy scraps of paper.

The ideas have been kept deliberately simple and open-ended and can be elaborated or extended as required. Many of these projects require little apparatus or complex preparation and some can be undertaken by children on holiday, without laboratory facilities or supervision.

The emphasis throughout is on conservation and minimum disturbance of the environment. For each section, a list of reference books and some background information is given.

Accurate identification often poses great difficulties to young students. This is not always helped by some popular reference books, which may give the impression that the bird, flower, insect, or seaweed found must be one of the relatively few that are illustrated or described in the text. Much can be accomplished by drawing, photographing or describing species accurately and giving them code numbers, until accurate identification can be made at a later date. Many of the projects deal with common species which cannot easily be confused with others.

Some students will need little or no more help than is given in the project outlines. Most will need some further help from the teacher in explaining points or in developing their own ideas to expand or modify the projects. Others will need to be taken step by step through the instructions if they are to make headway. All will need encouragement and frequent review of progress to be sure the scientific approach is maintained and that there is a continuing sense of achievement.

Standard methods, usually taught in schools or easily found out from books have not been given. Students should be encouraged to devise their own methods where this is feasible. Certain specialised methods or detailed procedures are given in the Appendix which should be consulted before starting any quantitative work, or wherever indication is given in the text.

To save time in searching for a suitable project, the following code has been used:

A   suitable for A level
O   suitable for O level or C.S.E.
E   suitable for younger age or non-examination groups

Where a letter is given in brackets, this indicates slightly less suitability for this level. Where a project is given as suitable for all levels it implies a much greater study in depth at A level with variation or extensions devised by the student.

This coding does not imply that the examiners of the various boards approve these projects at the different levels. Applications must be made in the usual way for approval of any project by an examining board.

At the time of writing the public examinations are under review at all levels and changes are likely in the 1980s. Presumably, optional projects will still form a part of many of the new syllabuses.

1 Seashore

Studies on the seashore lend themselves particularly well to holiday projects. More routine investigations may be carried out at Field Study Centres or at schools within easy reach of marine habitats.

Rocky shores in the South West of England, the Isles of Wight and Man, the Channel Islands, Wales and Scotland are probably the most rewarding. Some of the suggestions can be used or adapted for almost any type of shore, even those most influenced by man. Opportunities may also arise during holidays abroad.

Although seashore work has the advantage of yielding interesting results at all times of the year, the main difficulties involve times and seasons. To avoid disappointment, tide tables should be consulted to ensure the maximum use of the outgoing tide just before low water. This must be judged very accurately if the *Laminaria* zone of seaweeds is to be examined safely and adequately. A preliminary visit by the leader or teacher is recommended, especially when large groups are involved.

Safety is paramount and adequate information, warning and supervision must be provided, especially for younger children. The points to watch are edges and bases of cliffs, seaweed covered rocks and the incoming tide. The more interest created and the more absorbing the studies, the greater may be the risk of mishap or accident. On the other hand trouble may arise from any children who are not fully involved or are without a detailed programme of work. The dangers should not be over stressed, but adequate precautions must be taken and sensible clothing and footwear worn. Warm waterproof clothing is essential for most of the year in Britain. Except in summer, Wellingtons are useful, but can be treacherous on slippery rocks. When hot, beach shoes are better than bare feet, but if care is exercised most children manage well on rocks without shoes. A supply of waterproof dressings is the only first aid equipment likely to be needed.

Equipment should be kept to the minimum, especially when working on rocks or cliffs, as both hands are needed for balance and manipulation. Notebooks, pencils and polythene bags are usually needed by everyone and one or two penknives, plastic containers or pails with lids per group, plus measuring tapes and other essentials.

Seashore Projects

Cliffs

E	1	Bird observation (Short project)
OE	2	Bird observation (Longer project)
A(O)	3	Plant distribution related to physical features
A(O)	4	Plant colonisation of recent rockfalls
A(OE)	5	Comparison of cliff flora from different places
A(O)	6	Relation of plant species to springs
AO	7	Distribution of one plant species in relation to physical and biotic factors
OE	8	Effect of wind direction and force on tree and shrub growth

Rocky shores

E	9	Collection and identification of seaweeds
AO(E)	10	Zonation of seaweeds
OE	11	Comparison of floating, cast-up and growing seaweeds
AO	12	Zonation of animals
A(OE)	13	Animals associated with seaweeds
(A)OE	14	Distribution, movement and feeding in limpets
(A)OE	15	Distribution of barnacles
(A)OE	16	Distribution and colour variation in periwinkles
(A)OE	17	Distribution and mode of life of beadlet anemone
AOE	18	Study of rock pools

Sand and shingle beaches

A(OE)	19	Specialised flora of shingle beaches (Distribution and adaptations)
A(OE)	20	Specialised flora of sand dunes (Distribution and adaptations)
(A)OE	21	Distribution and habits of sand hopper
(A)OE	22	Distribution and habits of lug worm
(A)OE	23	The effect of man on beaches
	23.1	All types of rubbish
	23.2	Tar

Cliffs (Projects 1–8)

Cliffs are suitable for studies involving birds and plants. For bird watching, patience, binoculars and a good pocket reference book are essentials. It is of great help if a really knowledgeable person is available to lead the group. Often quite young boys meet this qualification, as well as or better than grown-ups, if ornithology happens to be their hobby. Such a person will know the best times and seasons to visit particular locations and what birds are likely to be seen. However all these matters can be discovered from books and much progress can be made by interested and observant youngsters, even without expert leadership.

1   E   Bird observation (Short project)

Using species lists obtained from the Royal Society for the Protection of Birds, tick off all those species identified during a half or whole day's cliff walk. Add notes as to what numbers are seen, their exact location (for example in air, on sea, on cliff face) and their behaviour (flying, gliding, using thermals, displaying, nesting, feeding young, and so on).

2   OE   Bird observation (Longer project)

If such a cliff walk is repeated in different weather conditions, at various seasons and in different locations, keeping accurate, detailed notes; a worthwhile long term project can be built up.

Razorbill *Alca torda* skull found on Cornish beach.

Interest will be increased by collecting such items as skulls, feathers, broken eggshells and bird of prey pellets, but it cannot be over emphasised that nests and eggs must not be touched or disturbed in any way. Local museums often have egg collections or stuffed birds which can be studied and drawn, as well as proving aids to identification and for studies of adaptation.

A great deal more can be learnt if it is possible to build a temporary hide or to visit a locality where such hides are provided for close-up views of birds. Photography of the birds and their habitats provides additional records of great value.

3   A(O)   Plant distribution related to physical features

Plants are best studied from the base of the cliff, as this is much safer than working from the top. Great care must be taken not to cause a cliff fall or to risk danger from falling rocks or stones. Cliffs should never be climbed alone and only after consultation with adults who are fully aware of local conditions. Warning notices should be heeded. Care should also be taken to avoid being cut off by a fast incoming tide.

Low cliffs suitable for project work with rocks covered with *Fucus* species in foreground.

Note taking and photography are better than taking specimens. If these are essential for identification, avoid duplication with other members of the group. Take enough of the plant, for example a length of stem, leaves, buds, flowers and fruits. DO NOT UPROOT ANY PLANTS NOR TAKE ANY SPECIMEN OF RARE OR PROTECTED PLANTS, but instead write very detailed descriptions. Cut specimens with pen-knife, tie numbered labels on plants to correspond with numbers in notebook and place in polythene bag or plant press. (See Appendix for use of press and mounting specimens.)

Make transects (see Appendix) from top to bottom of low sloping cliffs, if this is considered safe by adults or senior students. Carry this out in several representative places, at different times of year; many plants are recognisable only when in flower. Try to relate plant species to a) outcrops of different rocks,   b) degree of slope,   c) exposure to wind,   d) vertical and horizontal distance from high tide mark,   e) stone size, shingle and scree.

FIRST OF ALL TAKE CARE THERE ARE NOT LIKELY TO BE ANY MORE FALLS!

4   A(O)   *Plant colonisation of recent rockfalls*

Choose an area where there has been a recent rock fall or land slip. By using quadrat frames of various sizes (for instance half metre and decimetre) record the development of plant cover over several months. (See Appendix for detailed use of quadrat frame.) Do not remove plants from the quadrats for indentification but sketch them carefully or take similar plants from outside the area being studied. Include observations on growth of seedlings and trace their origin from nearby parent plants by vegetative means or seed dispersal.

Over a longer period, intra- and interspecific competition may be observed and studied in detail. For this, continue recording the changes in the vegetation within the quadrat frames for a year or more. When the bare ground is covered, if only one species is present, note the time when no further increase in numbers occurs and where the larger, stronger plants or those with a slightly better position overshadow the weaker ones. (This is known as intraspecific competition.) If more than one species is present, look for evidence that one of the species takes over dominance from another and what adaptations enable it to do so. (This is known as interspecific competition.)

5 A(OE) Comparison of cliff flora from different places
Projects 3 and 4 can be extended by comparing the results of
transects and/or quadrats from different rock types (such as chalk
and sandstone) for example from holidays in Sussex and Devon, or
chalk cliffs in different locations as in the south-east and north-east
of Britain.

Does the type of rock or the geographical location influence the
plants which you find to any great extent?

What other factors will you need to take into account before
coming to any conclusion?

6 A(O) Relation of plant species to springs
Freshwater springs are often seen at the junction of permeable and
impermeable rocks. The outlets of land drainage pipes may
sometimes be seen near the tops of low cliffs. In both places the flora
is likely to be very different from that of the drier parts of the cliff
face. It may include non-flowering plants such as algae (especially
Enteromorpha which shows a strong correlation with fresh water),
mosses and ferns as well as moisture-loving flowering plants.

Relate the distribution of such plants, as precisely as possible, to
relative humidity. This can be measured using cobalt chloride
paper or wet and dry thermometer (see Appendix for details).

Collect the water that seeps or drains out over a short or long
period of time by designing simple apparatus for the purpose, such
as a can tied to the end of a drainage pipe.

Test weigh samples of soil from wet and dry areas by drying
slowly until there is no further loss in weight. Thus calculate the
percentage of water held by the soil.

*7 AO Distribution of one plant species in relation to
physical and biotic factors*
The last four projects all involve identifying a small or larger
number of plants accurately. They are also long term projects of a
fair degree of difficulty. It may prove easier and more rewarding to
study the distribution of one typical abundant, well adapted and
readily identifiable cliff dwelling plant. Possible species include the
sea campion *Silene maritima*, bucksthorn plantain *Plantago
coronopus*, or thrift *Armeria maritima*. Here diagrams and accurate
scale sketch maps may be better than transects or quadrats.

Sea thrift *Armeria maritima* growing with lichens on low cliffs.

Relate distribution to vertical and horizontal distance from high tide mark, presence or absence of soil, percentage of salt in soil, exposure to prevailing winds, presence or absence of other plants, influence of man and any other physical or biotic factors which are judged important.

Study adaptations to avoid desiccation, for example reduction of leaf surface and water storage.

8   OE   *Effect of wind direction and force on tree and shrub growth*

Trees and shrubs growing on cliff tops often show one-sided growth. The reason for the uneven shape is that the buds most exposed to the prevailing winds do not develop. Investigate this in detail. Find out which trees and shrubs are most affected, in which positions and on which coasts.

Make careful drawings or take photographs to show the shape of the plants and to record leaf and flower bud development.

How far from the sea can this effect be seen in various coastal locations?

Wind force can be recorded by an anemometer (which can be purchased or made) and wind direction obtained by a weather vane. Reference to the Beaufort Scale also gives a good idea of wind force. (See John Sankey *A Guide to Field Biology*.)

Rocky shores (Projects 9–18)
9   E   *Collection and identification of seaweeds*
Seaweeds belong to the large group of simple, chlorophyll bearing plants, the algae. The green colour may be masked by other pigments as in the red and brown seaweeds.

Collect samples of red, green and brown seaweeds, following the advice given for collection of flowering plants on p. 8. Keep in large polythene bags away from sunlight until they can be examined, identified and possibly mounted, as described by Duddington (see references).

Two species of brown seaweed *Laminaria digitata* and *Himanthalia elongata* near low tide mark. *Fucus serratus* in foreground.

If you cannot identify them at once, classify according to   a) colour,   b) shape of plant body or thallus,   c) presence or absence of air bladders,   d) presence or absence of fertile portions containing reproductive organs.

Make a note of where each particular seaweed was found in relation to high and low tide mark, in rock pools or attached to other seaweeds.

10   AO(E)   Zonation of seaweeds

First carry out Project 9 to become familiar with the plants and the places where they grow.

Using information given in the Appendix, make transects from high to low watermark across seaweed covered rocks on exposed and sheltered shores. (Remember, it is the vertical distance from low water mark, not the horizontal distance, which determines the zonation of seaweeds.)

Laminaria digitata near low tide mark.

If there are problems with identification, give each unknown species a code number, for example, *green 1*, or *brown 2*, until identification is certain. Wrong identification is worse than incomplete identification.

The study may be simplified by studying brown seaweeds only, or *Fucus* species only.

11 OE Comparison of floating, cast-up and growing seaweeds

As the tide comes in collect the floating seaweeds, place them in polythene bags and compare with the species found on nearby rocks. This can be repeated with species found at high tide mark. Particularly interesting results may be expected after storms.

12 AO Zonation of animals

The transect described in Project 10 can be combined with an animal transect, or this may be done separately, or by another team operating at the same time. Only animals attached to rocks should be included.

This is a useful exercise if combined with a study of the adaptations of the animals to their position on the shore. It is also important for appreciation and understanding of classification into phyla, classes, genera and species if the age, interests and abilities of the students allow. It is essential in any A level study.

13 A(OE) Animals associated with seaweeds

Any survey involving seaweeds can be usefully extended to include the animals associated with them.

Collect specimens of whole seaweed plants, keeping each one in a separate large polythene bag, sealing it securely. Keep the bag in a cool place and turn it out at home or in the laboratory into a large bowl of sea-water. Wash each part of the weed carefully in the water and examine the animals in the seaweed and those which emerge from it. These are likely to include dog whelks *Nucella lapillus,* top shells *Gibbula* species, periwinkles *Littorina* species, and small crustacea and worms. Attached to the plants may be epiphytic seaweeds and colonial animals such as sea mats (Polyzoa), *Obelia* and other hydroids and sponges, as well as eggs of many species.

Identify and study each species of animal and count the numbers of each species. Relate the number to the total area of the thallus. This can be estimated by drawing round it on graph paper. Study camouflage and habits of the animals in temporary aquaria. Make drawings or take photographs. Return plants and animals to their original habitat as soon as possible or preserve in 2% formalin for further study.

Repeat collection at different states of the tide and at different

Brown seaweed *Ascophyllum nodosum* showing fertile portions, air bladders and the epiphytic red seaweed *Polysiphonia lanosa*. The green appearance of the thallus is due to the presence of chlorophyll incompletely masked by the brown pigment.

seasons. To simplify the study one easily recognisable species such as the bladderwrack *Fucus vesiculosus* can be selected and treated on its own, or compared with a closely related species such as the serrated wrack *Fucus serratus*, or a very different species such as knotted wrack *Ascophyllum nodosum*.

14 (A)OE   Distribution, movement and feeding in limpets
The common limpet, *Patella vulgaris*, is abundant and widely distributed. It is also easily identified.

Study distribution using both transects and quadrats (see Appendix) on different parts of the shore. Is distribution related to any of the following   a) degree of exposure to wind,   b) type of rock,   c) distance from high or low tide mark, bare slopes or sheltered crevices   d) presence or absence of plants or other animals?

Mark individual animals with waterproof paint and also around their shells on the rock using a different colour for each individual.

Return hours, days, and weeks later to study the distance moved by each animal in a chosen time.

Mark successive positions on graph paper.

Limpets and barnacles on rocky shore.

Study the path over which they have moved to see if there is any evidence of their removing algae from the surface with their rasping tongues.

15 (A)OE Distribution of barnacles

Study the distribution of barnacles in the same way and in relation to the same physical and biotic factors. Compare the distribution of the common species, *Balanus balanoides* with other species which may be present, for example *Chthamalus stellatus* which ranges higher up the beach. The occurrence of barnacles on the shells of other animals such as mussels and limpets is also worth investigating.

16 (A)OE Distribution and colour variation in periwinkles

Four species of periwinkle; *Littorina littoralis, L. saxatilis, L. neritoides* and *L. littorea* will probably be encountered in relation to transects and zonation, where their distribution in relation to high and low water mark is important. *Littorina littoralis* may be bright yellow, olive green, brown, black or striped. Study these colour variations and their distribution in a quantitative way and in relation to micro-habitat and predation. The dog whelk *Nucella lapidus* also shows considerable variation in colour and can be studied in a similar way. Collect empty shells to illustrate the colour variations.

17   (A)OE   Distribution and mode of life of beadlet anemone

Study the distribution of the common beadlet anemone (this is the reddish-brown jelly-like one) *Actinia equina* in relation to rock formation. Is it found in exposed or protected situations, on smooth surfaces or crevices, in rock pools or on rock surfaces? Make detailed maps and sketches to show where you find each animal in relation to these features.

Carry out experiments to find food preferences and times of feeding both in situ and in temporary aquaria.

To find out if anemones move from place to place, find some which you can mark around the base with waterproof emulsion paint. Make the circle a little wider than the animal so as not to touch it with the paint. Observe from time to time over a period of hours, days, or weeks to see if any movement has taken place.

18   AOE   Study of rock pools

These lend themselves to short or long term studies.

Choose one large pool with a well developed plant and animal community, or compare two or more pools of different sizes or at different positions on the shore.

Prepare accurate contour maps on squared paper. This is best done by measuring the shortest and longest diameters and drawing these in their correct positions. Then measure the circumference with a measuring tape or with string and a metre rule. Draw in the shape by eye as accurately as you can. Take soundings across a number of lines at decimetre intervals. Transfer the figures to the graph paper and draw in contour lines at 5 or 10 cm intervals. This map can also be used to construct a vertical section of the pool and to fill in the position of the main vegetation.

Record temperature at the surface and at the deepest part at different times of day and just before the tide covers the pool and just after the pool has become uncovered by the tide.

Record oxygen concentration at similar times using the modified

Winkler's technique given in the Appendix.

Take samples home to record carbon dioxide content (see p. 27) and salinity. Salinity can be determined by evaporating 50 cm³ to dryness and recording the weight of the dry salt and expressing this as a percentage. Salinity can also be accurately determined by hydrometer. Seal the end of a drinking straw and drop lead shot into it, then calibrate with known concentrations of saline solutions.

Visit the pool once a month over a period of a year if possible. As well as taking the physical and chemical readings each time, gradually compile a complete species list of plants and animals, with notes on their adaptations to microniches within this restricted habitat.

Sand and shingle beaches (Projects 19–23)
These have the advantage of easy access, with parts available for study at all states of the tide.

19 A(OE)   Specialised flora of shingle beaches
Some shingle beaches have a specialised flora, notably Dungeness in Kent and Chesil Beach in Dorset. Some parts of these have been designated as Nature Reserves where plants may not be picked or

Silverweed *Potentilla anserina* on shingle beach. (Uig, Skye)

disturbed. In any case the plants are better studied in situ and samples should rarely, if ever, be removed from the habitat.

Make species lists and study distribution in relation to distance from high tide mark, shelter and stability of the substrate (whether the stones are constantly shifting position or not).

Study adaptations to water conservation such as reduction in leaf surface, hairy leaves and succulent leaves and stems. (*N.B. These adaptations may be related to physiological rather than actual drought. This is caused by the high salinity of the substrate making it difficult for plants to obtain water by osmosis.*)

Sea plantain *Plantago maritima* on shingle beach. (Uig, Skye)

Observe and record reproductive adaptations, such as rapid spread by means of runners or other vegetative structures.

Study pollination devices in relation to the specialised insect fauna or the agency of wind.

Dispersal of seeds in relation to the environment may also be studied.

20   A(OE)   Specialised flora of sand dunes

Sand dunes also have a specialised flora and fauna and may be studied along the same lines as given above. In addition the particular adaptations of marram grass *Ammophila arenaria* and its role in dune stabilisation make an excellent autecological study.

Choose an area of the dunes which contain some bare sand adjacent to an area colonised by marram grass, say about 10 km square. Represent the area covered by marram and the bare area as accurately as you can, indicating the density of the marram by differences in shading.

Devise some way of measuring the stability of the substrate in a) the completely bare sand, b) the partially colonised area and c) where the marram is thickest. One method might be to see how deep a footprint sinks in when walking over the three areas.

Choose half metre squares on the edge of the marram grass. Mark one tuft of the marram grass in a distinctive and permanent way, for example by tying coloured string low down where it is not likely to be spotted and removed by others. Mark the position of this tuft on the graph paper and indicate the presence of other tufts in the quadrat. Repeat at monthly intervals, orientating the quadrats with the help of the marked tufts. So record the growth of the marram grass in each area. Make drawings to show the way in which the creeping stem roots at the nodes and thus binds the grass and spreads to new areas.

Take leaf samples back to base and cut transverse sections to show the way in which the leaf is rolled and the stomata are protected from excessive transpiration. Study and write notes on any other features which you consider are adaptations of the plant to its environment.

21 (A)OE *Distribution and habits of the sandhopper* Talitrus saltator

Around high tide mark on a sandy shore, this small crustacean may be found amongst heaps of washed up seaweeds. Also look for the small holes which denote its presence in the wet sand. By a quick digging action with the hand, the sandhopper can be scooped up quite easily from a few centimetres below the surface. Study the animal briefly and record the time it takes to bury itself again and the method it uses.

Carry out quantitative work on its distribution in relation to the humidity of the sand, distance from high tide mark and depth to which the sandhopper buries itself. For the humidity, take samples

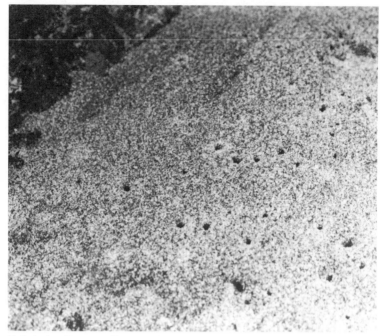

Exit holes from sandhopper burrows.

of sand from each area in which sandhoppers are found, placing each sample in a separate labelled polythene bag. Weigh each sample and dry slowly until there is no further loss in weight. Express the amount of water present as a percentage and find the range of water content of the various samples. Compare this percentage with that found in apparently dry, and very wet sand.

22   *(A)OE   Distribution and habits of lug worm* Arenicola marina

Lower down the shore the holes and casts of the lug worm will almost certainly be found. These worms are not easy to dig up, but a lesson in this art may be readily obtained by watching the technique of the bait diggers. If you are unable to achieve similar results by lack of a suitable spade or expertise, a tactful request may result in the gift of a complete specimen, in which you can study the adaptations to breathing and burrowing. A good description of these processes is given in C. Yonge *The Seashore*. Use quadrats and transects (see Appendix) to find the distribution of the worm in relation to high and low tide, exposure of the shore, type of sand (soft, firm, ribbed, coarse, fine, percentage of humus content).

23   *(A)OE   The effect of man on beaches*
Any type of shore line or beach offers interesting possibilities for this project.

23.1   *All types of rubbish*
Survey several small beaches, or one beach divided into sections. Record and/or collect human artefacts, dividing them into various categories: for example flotsam and jetsam brought in by the tide, including fruit and vegetable remains, bottles, plastic objects, tar and oil deposits, rubbish deposited by cliff dwellers, rubbish left by picnickers. Repeat at various seasons, making comparisons of the results and suggesting reasons for the differences.

If the amount is not too much, or if the team is large, collect the rubbish after first recording it by means of maps, photographs and notes. If one or more sections are completely cleared they can be inspected again at intervals to see how fast litter accumulates. Be sure no-one else is collecting rubbish on this part of the shore. If they are, take these collections, by the Council for instance, into account. If you are working on a large scale, inform and co-operate with the Local Authority who may welcome your help and provide extra disposal facilities. BEWARE OF BROKEN GLASS AND TAR.

Try to answer these questions and then make up others of your own:

To what extent do you consider rubbish on the beach   a) spoils its appearance,   b) is dangerous,   c) affects other living organisms?

What can you learn from the distribution of rubbish on the beach?

Can you suggest any long term solutions?

What sort of results do you get after storms?

23.2   *Tar*
Grade sections of the beach according to the amount of tar deposit on a five point scale. This scale can be adapted to suit local needs.

0   No tar found after half an hour (or longer) inspection
1   Some tar found after careful searching over same length of time (one or two patches only)
2   Tar found in small infrequent patches
3   Tar never far out of sight
4   Difficult to step without treading on tar
5   Beach covered by oil slick

Try to find out the origin of the tar, how long it has been there, whether any steps have been taken to get rid of it, whether it has any effects on living organisms, what local people and visitors think of it, and whether the distribution of the tar can be related to shape of coast and direction of prevailing winds. Here again the Local Authority may prove both informative and helpful. Always send stamped addressed envelopes and a courteous letter when writing for information and state the object of your enquiries.

Further Reading

1   Angel, Heather 'The Seashore', *Photographing Nature* Fountain Press 1975
2   Barrett, J. *Life on the Seashore* Collins 1974
3   Barrett, J. & Yonge, C. M. *Collins Guide to the Seashore* Collins 1958
4   Brightman & Nicholson, *The Oxford Book of Flowerless Plants* O.U.P. 1966
5   Campbell, A. C. *Guide to the Seashore and Shallow Seas of Britain and Europe* Hamlyn 1976
6   Chinery, Michael *The Family Naturalist* Macdonald & Janes 1977
7   Dickinson, C. I. *British Seaweeds* Eyre & Spottiswoode 1963
8   Duddington, C. *Beginners' Guide to Seaweeds* Pelham Books 1971
9   Dance, S. P. *Seashells* Hamlyn 1972
10   Evans, I. O. *Observers Book of the Sea and Shore* Warne 1962
11   Felix, J. *A Colour Guide to Familiar Sea and Coastal Birds, Eggs and Nests* Octopus Books 1976
12   Gibson-Hill, C. A. *A Guide to the Birds of the Coast* Constable 1976
13   Hepburn, I. *Flowers of the Coast* New Naturalist Series No. 24 Collins 1954
14   Ingle, R. *A Guide to the Seashore* Hamlyn 1972
15   Lewis, J. R. *The Ecology of Rocky Shores* E.U.P. 1964
16   Sankey, J. *A Guide to Field Biology* Longman 1958

17 Saunders, D *Birdwatching* Hamlyn 1975
18 Shell Education News *On the Beach*
19 Soper, Tony *The Shell Book of Beachcombing* David and Charles 1972
20 Step, Edward *Shell Life* Warne 1945
21 Vevers, G. *Seashore Life* Blandford Press 1969
22 Yonge, C. *The Seashore* Fontana New Naturalist 1963

2 Freshwater

Like the marine habitat, freshwater abounds in excellent material for investigating the structure of an ecosystem, the principles of classification, variation, and adaptation to the environment.

Freshwater localities can be visited at any time of day and at most times of the year. There are few schools which are not within easy reach of such a habitat, even if it is only a park lake or a school pond. Freshwater aquaria are much easier to maintain than marine ones. The physical dangers are not great, but sensible precautions should be taken, especially with young children or if boats are used.

Small rivers or streams, drainage ditches and ponds are most suitable for projects, especially if they are unpolluted. The effect of pollution on freshwater habitats forms an extensive project of its own.

Permission must be obtained for working on private land. This is usually readily given if the owner is assured of the minimum environmental impact. The country code must be scrupulously observed. Particular care should be taken not to leave behind or 'lose' polythene bags which can be dangerous to livestock if swallowed. The habitat should be left as it was found and animals and plants taken for study purposes should be returned on the next visit.

There are many excellent background and reference books. The keys published by the Freshwater Biological Association are especially recommended for advanced work. Most students will need help using the keys at first.

Freshwater Projects

Rivers and large streams
A(O)   24   General survey of freshwater ecosytem in limited stretch of river or stream
A(O)   25   Making the study quantitative
A(O)   26   Adaptations to aquatic life
A(O)   27   Study of group of aquatic animals
A(O)   28   Autecological (one species) studies
       28.1   The yellow water lily
       28.2   The great pond snail
AOE   29   Projects concerned with pollution
(A)OE   30   Small stream survey

Drainage ditches
A   31   Colonisation of newly dredged ditches
A   32   Comparison of 2 or more contrasting channels or ditches
A   33   Estimating sampling efficiency of different methods of collecting aquatic animals

Ponds, lakes and reservoirs
OE   34   Ponds
AO   35   Succession
(A)OE   36   Constructing a school pond
(A)OE   37   'Save the Village Pond'
A   38   Comparing flora and fauna of sheltered and exposed shores of lakes
A   39   Comparing flora and fauna of different lakes
A   40   Field experiments using artificial vegetation
A   41   Colonisation of new reservoirs
A   42   Effect of herbicides
(A)OE   43   Bird watching and recording

Rivers and large streams (Projects 24–30)

Many of the projects described have been carried out by A level students on the River Beult in Kent.

24   A(O)   *General survey of freshwater ecosystem in limited stretch of river or stream*

Choose a readily accessible part of the river and sample it first to check that it contains sufficient material of interest. A net swept two or three times amongst vegetation and emptied into a large dish should contain a considerable number and variety of animals.

It is best to start the study between May and October when vegetation is abundant. If possible, choose a region which shows marked contrasts in physical features, such as a steep and shallow bank, above and below a weir, shaded and non-shaded, fast and slow flowing, and so on.

PHYSICAL FEATURES

Equipment needed includes a small dinghy or similar boat if possible, a large scale map (at least 1:10,560), clip board with squared paper, ruler, pencil, rubber, 50 metre measuring tape, metre rule, stop watch, Secchi disc (for measuring turbidity), environmental comparator with light and temperature probes (or thermometer), two stout poles or canes, length of fairly thick nylon cord, small trowel and strong, medium sized polythene bags.

Make an accurate scale map based on a large scale Ordnance Survey Plan or Map using the clip board and squared paper. The width of the river can be measured at a bridge, but if a boat can be used, make several transects at points of interest. Fix a cane or pole on each bank and the nylon cord between them. Take soundings of the depth at 2 or 5 metre intervals, using the metre rule or longer pole marked in decimetres. Take temperature and light readings at the same places, at the surface, mid-depth and bottom. Determine the decrease in light intensity with depth using a Secchi disc. This is a white disc with a diameter of 25 cm which is allowed to sink down slowly on a marked line until its outlines just disappear. The depth at which this happens is the depth of visibility and this is a measure of the transparency of the water. It is very much affected by the plankton content of the water, amongst other factors.

Estimate the rate of flow at several places, using an orange, if this is retrievable, or a piece of stick. Take three readings from several places, for example on both banks at a bend, and at a straighter part, at the bank and in the centre, and so on. Use a stop watch and calculate the mean of these readings.

Collect samples of the substrate on the bank and on the river bed and place in labelled polythene bags. Write the labels clearly in pencil on rough paper and enclose with sample, tying it securely. If the substrate is fine, put it through a soil sieve on return to base and estimate the proportion of different sized particles.

CHEMICAL FEATURES

The percentage of dissolved oxygen may be sampled by means of a portable probe, but this is very expensive. Details of a modified Winkler's technique which can be used in the field is given in the Appendix. The percentage of carbon dioxide can be estimated by titration with sodium hydroxide solution in the laboratory using phenolphthalein as an indicator. Tablets for estimating hardness of water, due to calcium, are available from biological suppliers and are very simple to use. Determine the pH of water with universal indicator or paper or with a pH meter.

Emergent vegetation, Common bulrush *Schoenoplectus lacustris*.

Water samples should be taken in various places and at different depths. Transport them in clean plastic bottles, which must be completely filled. Carry out analysis as soon after collection as possible.

PLANTS

The aquatic vegetation may be classified as   a) emergent (rooted in mud but growing above water surface) for example bulrush *Schoenoplectus lacustris*,   b) submerged, such as Canadian pondweed *Elodea canadensis* or   c) floating or partly floating, as duckweed *Lemna* species.

The rules about collecting plants, given on p. 8 should be observed. Cut emergent plants close to the water surface with a sharp knife and label immediately, using pencil and tie-on or push through label, giving site information and other details. Collect floating and submerged plants from boat or from banks using a plant grab or an old garden rake. The grab can be bought or made and consists of a lead filled cylinder with three strong hooks embedded in it and tied to a strong cord. These grabs are often lost by inexperienced or over enthusiastic throwers and for this reason a rake, though not quite so effective, may prove cheaper. All plant samples should be enclosed in polythene bags after labelling.

Emergent vegetation, common bur-reed *Sparganium erectum*.

Emergent vegetation, Water plantain *Alisma plantago-aquatica*.

ANIMALS
Equipment: plankton and sweep nets, large pie dishes or developing dishes, plastic spoons, large plastic jars with screw on or press on tops for transporting animals, large polythene containers for transporting water, paper and pencil for labels.

Plant and animal plankton may be obtained by a special fine meshed plankton net with a small bottle attached. Pull this behind the boat on a cord or use on a handle from the bank. Use nets of a coarser mesh to obtain the larger animals. Sweep the net through the open water, amongst the weeds, or near the bank and empty the contents into a large dish for examination. Use a plastic spoon to remove animals for further study into large open neck plastic jars. These should not be completely filled with water as many aquatic animals are air breathing. Take a supply of clean river water in large containers. Collect samples of stones and substrate to search for animals later, enclosing them with labels in polythene bags. As much identification as possible should be done in the field to avoid harm to organisms during transport.

On return to base, remove all plants and animals immediately from their containers and place with river water in large dishes or

Examining the catch from a plankton net which has a small glass or plastic
tube attachment for retention of small animals.

bowls. Pencil labels do not last long in water and so make them more
permanent and firmly attached. Cover dishes containing animals to
prevent their escape and the evaporation of water. Identify and
name plants, study them and draw them.

Identify animals and place different species in smaller covered
dishes. Arrange the dishes in phyla, classes, orders, families and
genera. Also classify animals according to their position in the river,
for example, surface film, open water, on vegetation, under stones,
in mud, and so on. Petri dishes with lids are useful temporary
aquaria for smaller animals. Keep them out of the sun and place the
animals in larger aquaria or return them to their habitat as soon as
possible. Many animals will be found in the dishes containing the
vegetation. Rinse this a little at a time in clean river water to remove
animals. Many will be found clinging to the sides and bottom of the
dish when the plants are removed and the mud has settled.

25   A(O)   *Making the study quantitative*
After attempting to answer the questions *What?* and *Where?* and
compiling a species list of plants and animals it is instructive to ask
How Many?

A net can be used in a semi-quantitative way by making measured sweeps, say three metres, emptying the contents into a dish and counting the animals. The results become more meaningful by comparing sweeps at different stations, for example, mid stream, left bank, right bank, in open water and amongst vegetation. Repeat them at different seasons. An alternative method is to collect for a certain length of time, say, half an hour, and count the number of animals collected. Both methods will give more accurate comparisons if carried out by the same person on each occasion. If this is not possible, standardise the methods very carefully. Various methods can be devised to check the accuracy and value of such quantitative methods of sampling. See p. 38.

26   A(O)   Adaptations to aquatic life

Many further projects may be devised to answer the question, *How?* How are the plants and animals adapted to their particular place in the environment, such as, the water lily and the water skater to the surface film, or the Canadian pondweed and the water boatman to the open water?

The adaptation of animals for breathing, movement, feeding and reproduction offer great scope for investigation to suit individual interests. Information and suggestions are to be found in such books as H. Mellanby *Animal Life in Fresh Water*, A. D. Imms *Insect Natural History* and J. Sankey *A Guide to Field Biology*. Original notes and drawings are of great value in such a project, those copied from books are not.

27   A(O)   Study of a group of aquatic animals, such as, snails, flatworms, crustacea or insects

The methods used for collecting, counting and observation are described in the previous projects. Snails are a particularly rewarding group to choose as they are large, limited in number and have fairly easily identified species. The F.B.A. key is a great help. Be sure to follow the key and read the notes as well as looking at the drawings. The best time for study is September and October when the snails are nearly full grown, or in May and June when they are full grown (annual species) and spawning. Most snails are annuals, dying soon after egg laying.

Snail's eggs found on leaves of water plants are most interesting to view under the hand lens or the microscope. Hatch them out in covered petri dishes, renewing the river water from time to time and transferring the young snails with a fine paint brush to larger aquaria after a week or two. Carry out experiments on feeding and rate of growth of the young animals.

28 A(O) Autecological (one species) studies
These involve the study of the relationship of one species to its environment.

28.1 The yellow water lily
Nuphar lutea is an excellent plant to choose if sufficiently abundant locally. Include distribution maps, quadrats and drawings to show morphological and anatomical features that are adaptations to aquatic life, such as the long petiole and the air storing tissues. Include the rhizome with its capacity for food storage and vegetative reproduction.

Floating vegetation, Yellow water lily *Nuphar lutea*.

Many animals make use of the plant for food, shelter and a reproductive site. If possible, collect five leaves at random (see Appendix) in each month from May to October. Place each leaf in a separate, labelled and secured polythene bag. On return, remove all animals and eggs carefully from the leaf and place in a petri dish with river water. Identify, count and record the animals and eggs. Trace round each leaf on graph paper, including all holes and damaged sections. Calculate the area of the leaf and the area eaten or damaged. Try to explain your results in terms of the use made of the leaf by the animals throughout the season. Observe insects pollinating the flowers, date of flowering, fruiting and seed dispersal.

The suggestions given above could be adapted for other species of plants.

28.2 The great pond snail

Lymnaea stagnalis or another abundant, readily identified species could be chosen. *L. stagnalis* is a perennial snail, living for two or three years. Make a collection in late autumn or spring and sort the animals into size groups according to age. By means of bar graphs get some idea of the structure of the population.

Calculate the potential and actual reproductive capacity under laboratory conditions by keeping a few mature snails in a large aquarium. Count the number of egg masses and eggs laid in a season. See how many snails survive to lay eggs themselves. Try to devise ways of measuring this in the field.

As the snails are kept under observation many other experiments will suggest themselves, relating to feeding, growth and reproduction. While best kept in a large aquarium, they can be kept for short periods of time in a model stream such as that described in *Nuffield Biology*, Book 3. Sections can be prepared with the same substrate and different vegetation or no vegetation, and with different substrates and the same vegetation. Put in about twenty snails and test their preferences, if any.

29 AOE Projects concerned with pollution

Early in 1971, the Advisory Centre for Education, the *Sunday Times* and the Nature Conservancy set up a young people's investigation into water pollution in Britain, which was well received and supported. Full details of this survey are given in Richard Mabey *The Pollution Handbook*.

On the whole the methods given are most suited to younger children, but some of the ideas, especially those concerning indicator organisms are valuable at more advanced levels.

30   (A)OE   Small stream survey

This may be more rewarding than tackling a small section of a large stream. A small stream here means one a metre or so wide. It has a smaller community of plants and animals, with fewer problems of identification. The physical and chemical factors can be monitored more thoroughly. The following method was found to work well in a small woodland stream in Kent.

Clear out any artefacts (man-made, introduced objects). Dispose of these safely and suitably and keep a list of those removed each week. Before disposal, examine some of them to see if they have been colonised by plants or animals. If necessary bring them back to base for identification of these organisms.

Make an accurate scale map of a 10 m stretch of the stream, on as large a scale as possible. It helps if this can be based on a large scale Ordnance Survey map.

Measure the width and depth of the stream at metre intervals and record the temperature and rate of flow at regular intervals (at least monthly). Keep weather records and relate to other data.

Collect samples of all plants in the stream or on the banks and all animals in the stream (see notes on p. 31 about collecting). Write notes on seasonal changes such as flowering, fruiting, leaf fall and so on. Return animals after identification. Animals can be caught using small metal or nylon flour sieves rather than nets. Turn over stones and examine them carefully for leeches, flatworms, snails and insects then replace them the right way up. Wash samples of the substrate in the sieve and examine the washings for animals in the laboratory.

Make transects across the stream at one or two metre intervals filling in the positions of the plants where possible.

Introduce plants such as duckweed *Lemna* species and Canadian pondweed *Elodea* in measured amounts and counted animals such as snails, water shrimps, waterlice (if these are not already present). Search for the animals and plants and recount at intervals to see if they have been able to establish themselves.

Analyse substrate and soil on bank for size of particles, percentage of humus and presence of minerals, using methods learnt in Biology classes or devised specially.

Keep accurate field notes and include tables, maps, diagrams and drawings from life.

N.B. This work is of greater value and interest as a group or class project with adjacent stretches of the stream adopted by pairs of students, to gain a picture of the life of the whole stream.

Drainage ditches (Projects 31–33)

Low lying lands with an extensive system of dykes or drainage ditches, as in East Anglia, Sedgemoor, and the Pevensey and Romney Levels, are very rich in aquatic flora and fauna. These areas form valuable refuges for some rare species, but they are fast diminishing owing to building and pumping schemes. It is important to collect very sparingly, returning all species to their original habitat. Schools might well attempt fairly ambitious surveys and get the material published in order to increase the chance of conservation of rare species. The Royal Society under the auspices of its Scientific Research in Schools supported such a survey on the Pevensey Levels in which several local schools took part from 1968–72. At the time of writing, news of an intended Nature Reserve on the Levels has been published in the national press.

A typical drainage channel on the Pevensey Levels.

Many of the projects described for rivers can also be used in this habitat. In the Pevensey Levels in Sussex, for example there are over three hundred miles of easily accessible waterways within a relatively small area. There is plenty of scope for well organised projects with conservation in mind.

31   A.   *Colonisation of newly dredged ditches*

On the Pevensey Levels, it is possible by observation, or consultation with the Sussex River Board, to find out when ditches are being dredged. The main channels are dredged about every five years and the minor ones much less frequently by the farmers.

Excavator at work on Pevensey Level drainage channel.

By arriving at the right time, the ditches may be sampled by sweeps of the net and by collecting vegetation just ahead of the dredger and then at monthly intervals after dredging.

Collect and record systematically. If time is limited, concentrate on certain species or groups of species for example gastropods (snails). The rate of colonisation probably depends on factors such as season of dredging, extent of dredging, width of channel, position in relation to other channels and so on. Although this is a long term project, it can be taken on by successive groups of

Drainage channel many years after last clearance. (Pevensey Levels)

students and the pattern which emerges will be full of interest and brimming over with other questions to be answered.

32   A.   Comparison of two or more contrasting channels or ditches

This could be a very short project taking half a day, or it can easily be extended over a whole year.

For a short term project, collect plants and animals by all means available (see p. 28) in equal lengths of channels that show obvious physical and vegetational differences. Collect for a chosen period of say half an hour, or an hour. Identify and count the organisms. Return them to the habitat as soon as possible. Try to relate differences in flora and fauna to measured physical features such as width, depth, exposure to wind, nearness to sea, and period of time since last cleared.

For a longer project, repeat the procedure at intervals throughout the summer or the year.

If there are too many identification problems, concentrate on one group of animals such as larger crustaceans, water bugs or snails.

33 A. Estimating the sampling efficiency of various methods of collecting aquatic animals

Aquatic habitats are notoriously difficult to sample in a quantitative way, owing to the patchy or non-random distribution of animals.

Choose a length of channel which is varied in vegetation and contains at least one end section (near a farm gate, for instance) where vegetation and animals tend to collect. Choose a station at each end of the channel and one in the middle.

Make a series of one metre sweeps as near to the bank as possible, holding the net so that it is just below the surface of the water. Empty each catch into a separate dish or collecting bottle. Identify and count the animals collected. Repeat, using three metre sweeps at each station. Repeat a second time holding the net as far out in the channel as possible. Repeat a third time using nets of different mesh size, using different operators and sweeping at the bottom of the channel. Fill standard sized polythene bags, each with a different type of vegetation. Empty out the plants into separate large bowls with some filtered water from the ditch, or some tap water. Rinse small lengths of the vegetation in clean tap water or rain water and count the animals which emerge and those left in the bowl afterwards.

It is important to carry out these sampling techniques at the same time of day, at the same time of year, and in the same weather conditions. Analyse the results graphically and statistically in regard to both numbers of species and individuals. What do you consider would be the best sampling programme for a future detailed survey, in the light of your results?

Ponds, lakes and reservoirs (Projects 34–43)

Many of the methods given above can be adapted for these habitats. Here are a few additional ideas for each locality.

34 OE Ponds

If relatively small and shallow as well as being interesting biologically, ponds are much better habitats for young children to study than rivers, streams or ditches. There is often a pond in the school grounds, or within easy reach. Supervision is easier and dangers less. Quite small children get enormous enjoyment out of pond study and can learn a good deal by scaling down some of the above mentioned methods to their interests and capabilities. Pond dipping is a well established feature of nature study which need not be described further here.

35 AO Succession
Older students can learn about plant zonation and succession from a pond. This is of particular value if maps are made from year to year showing the changes. When the hydrosere (pond) is in danger of becoming a xerosere (dry land), the clearing out and revitalising of the pond can make another project. If the pond is large, part can be allowed to follow the natural succession while part is prevented from silting up and the results can be compared.

36 (A)OE Constructing a school pond
Details of procedure are outside the scope of this book but are well described elsewhere (see Further Reading, p. 41).

Colonisation by organisms from outside and the fate of artificially introduced plants and animals would make suitable individual or group projects at any level.

37 (A)OE 'Save the Village Pond'
This project has been well publicised in the U.K. and details of methods used are still obtainable from the *Telegraph*. Any fieldwork project that benefits the community, as well as being sound ecology, has very much to recommend it. Lakes may be man-made or natural features, obviously not sharply distinct from ponds, except that they are larger. They lend themselves to more ambitious studies, of which only a few can be outlined here.

38 A. Comparing flora and fauna of exposed and sheltered shores of lakes
Using some or all of the methods of sampling given on p. 28 compare sheltered and exposed shores of lakes in relation to temperature, wave action, substrate, flora and fauna.

39   A.   Comparing the flora and fauna of different lakes

In a region where there are many lakes of various sizes and types, as in the Lake District, North Wales or Scotland, compare the insect, gastropod (snail), flatworm or crustacean fauna of as many lakes as possible. One way of doing this is to choose one or more stations on each lake and collect, using the most appropriate of the methods described above (p. 28) for a standard length of time. Compare the results obtained from each lake and attempt to relate them to physical features such as area of lake, altitude and so on.

40   A.   Field experiments using artificial vegetation

If a private stretch of river or pond or lake is available, much interesting work can be done on colonisation by introducing artificial vegetation. This can be obtained from aquarium dealers or made from suitable plastic material. Sheets of plastic or glass slides can be used for similar purposes. These must all be firmly secured, so that they can easily be retrieved at intervals and examined for the presence of snails, flatworms and epiphytic algae. Further details of such methods are given in J. Schwoerbel *Methods of Hydrobiology (Freshwater)*.

Some snails actually seem to prefer artificial vegetation to natural plant life and this aspect might be investigated fully.

41   A.   Colonisation of new reservoirs

If you are fortunate enough to be within easy reach of a newly constructed private or public reservoir and can get permission to work there, colonisation may be studied by adapting the methods given for drainage ditches. Study colonisation by plants using half metre quadrats constructed of wood, with string marking the decimetre squares. These readily float on water and percentage cover by floating plants can be measured at intervals.

42   A.   Effect of herbicides

Many reservoirs, lakes, rivers and drainage ditches are treated with herbicides by the water authorities. Details of treatment can be obtained from the appropriate authority and effects on plant and animal life studied over a period of time. Permission must be obtained and co-operation requested.

43 (A)OE Bird watching and recording

All the aquatic habitats dealt with here offer opportunities for bird watching and related projects. Many such water bodies have areas designated as Nature Reserves, some of which are open to the public and others for which permission must be obtained to visit. It may be possible to carry out some of these projects under such conditions, but the emphasis must be on observation rather than collection. For projects involving birds see p. 6 and the books in the further reading sections, especially D. Saunders *Bird Watching*, Chapter 1.

Further Reading

1   Allen, G. R. and Denslow, J. B. *Freshwater Animals* O.U.P. 1970
2   Bennet, G. W. *Management of Lakes and Ponds* Van Nostrand 1971
3   Burton, Robert *Ponds, Their Wildlife and Upkeep* David and Charles 1977
4   Clegg, J. *The Freshwater Life of the British Isles* Warne 1974
5   Clegg, J. *The Observers Book of Pond Life* Warne 1956
6   Clegg, J. (Ed.) *Pond and Stream Life* Blandford 1973
7   Engelhardt, W. and Merxmuller, H. *The Young Specialist Looks at Pond Life* Burke 1964
8   Haslam, S., Sinker, C. and Wolseley, P. *British Water Plants* Reprint from Field Studies Vol. 4, 243–351 (1975)
9   Hynes, H. B. *The Biology of Polluted Waters* Liverpool University Press 1963
10 Imms, A. D. *Insect Natural History* Collins 1947
11 Leadley Brown, A. *Ecology of Fresh Water* Heinemann 1971
12 Mabey, Richard *The Pollution Handbook* Penguin Education 1974
13 Macan, T. T. *Freshwater Ecology* 2nd edition Longman 1973
14 Macan, T. T. and Worthington, E. B. *Life in Lakes and Rivers* Collins 1968 and Fontana Press 1972
15 Macan, T. T. *A Guide to Freshwater Invertebrate Animals* Longman 1959
16 Macan, T. T. *Biological Studies of the English Lakes* Longman 1970
17 Mellanby, H. *Animal Life in Fresh Water* Methuen 1963, also Science Paperbacks Chapman & Hall 1975
18 Nuffield *Key to Pond Organisms* Longman 1970
19 The Nuffield Foundation *The Maintenance of Life* (Text III) Longman 1966 N.B. This series has now been revised and the relevant information may appear in different volumes.

20   Schwoerbel, J. *Methods of Hydrobiology (Freshwater)* Pergamon Press 1970
21   Southwood and Leston *Land and Water Bugs of the British Isles* Warne 1959
22   Ward, H. B. and Whipple, G. C. *Freshwater Biology* Second edition, revised by W. T. Edmondson and John Wiley New York 1959

See also titles on the Chapter 1 Further Reading List, Nos. 6, 16 and 17.

Series of Keys published by the Freshwater Biological Association:
No 5   *British Species of Freshwater Cladocera*
    13   *British Fresh and Brackish Water Gastropods*
    14   *British Freshwater Leeches*
    16   *Revised Key to British Water Bugs*
    18   *British Freshwater Cyclopid and Calanoid Copepods*
    20   *Nymphs of British Species of Ephemeroptera*
    23   *British Species of Fresh Triclads*
    25   *Statistical analysis of samples of Benthic Invertebrates*

3 Mountain and Moorland

British mountain and moorland is of special interest to the biologist as it shows the dependence of organisms upon their environment on a large scale. It includes a whole range of habitats with restricted and often much specialised flora and fauna. Much of it now lies within the National and Forest Parks, many of which are near large centres of population. Twenty million people, for example, live within fifty miles of the borders of the Peak National Park. All the Parks have excellent information centres and produce valuable educational material. Many schools already make considerable use of such resources, visiting nature trails as well as day and residential centres for more detailed studies.

Project work, if properly planned and prepared, can add greatly to the understanding and enjoyment of mountain habitats. Some of the projects in other sections of the book can easily be adapted to this environment, particularly numbers 30, 38, 39, 60, 64 and 82.

This chapter outlines projects useful for schools which have not yet developed their own programmes and for groups, or individuals, on field excursions or holidays. Some of them might provide biological investigations for students at outdoor pursuits centres. ·

Those venturing into remote country, whether in twos or threes or in larger parties, should wear correct clothing and take safety precautions. Full information may be obtained from any of the National Park Information Centres. Briefly summarised the main points are:

Never go out alone
Never go out in severe weather
Always inform someone of your route and proposed time of return
Wear warm, waterproof clothing and strong footwear (boots or walking shoes)

Mountain and Moorland Projects

44 A(O) Distribution of the flatworms Crenobia
(Planaria) alpina *and* Polycelis (Planaria) nigra *in mountain*
streams

This is a suitable project for any part of the country where
mountain streams are accessible both above and below 2,000 feet. It
is best carried out in summer on warm days.

The equipment required includes an Ordnance Survey map
(scale 1:25,000), a centigrade thermometer, a stop watch or watch
with second hand, a hand lens on a string round the neck, about
ten polythene bags with labels and ties, a small dish and clean paint
brush, a 100 m measuring tape.

Crenobia alpina has awl-shaped tentacles of varying length
situated at the sides of the 'head' which has two eyes. The dorsal
colour is grey to black, less commonly brownish, while the ventral
surface is paler. *Polycelis nigra* is black or brown with no visible eyes
(except under the microscope when the numerous eye spots may be
seen) and a triangular 'head'. The ventral surface is not lighter.
Both species are found under stones, but the former is said not to
occur where the water temperature rises above 15°C for any length
of time. It is characteristic of spring heads and small cool, swiftly
flowing streams. The latter species occurs in waters up to 20°C.

Others species of flatworms that may be encountered in
mountain streams are *Polycelis felina* and *Phagocata vitta* and they
should also be recorded.

With the help of the map and personal investigation, choose a
mountain stream which you can follow from above 2,000 ft for a
considerable length of its downward course. Work out from the
map the distance in metres between the 100 ft contour lines which
are crossed by the stream. Starting at the highest possible level and
at a known position on the map, turn over several stones in the
stream until you are able to identify the flatworms which are slug-
like and about 5 mm long. Brush several of them off carefully with a
clean paint brush into the small dish containing some of the stream
water. Replace all stones the right way up. If possible, identify the
worms in the field, but if this is not possible enclose a few in a
polythene bag with a small stone and some of the stream water,
keeping them out of the sun. Record the temperature of the water
and the rate of flow. It may not be easy to record rate of flow in a
mountain stream but consult Project 24 and devise as accurate a

Mountain stream.

method as possible. Measure the distance on the ground to the next 100 ft contour line and check with any landmarks. Repeat the procedure at each 100 ft interval.

If you find sufficient worms, it may be possible to make the study quantitative by counting or collecting the worms from ten or more similar sized stones at each station. If you find no worms, collect and identify the animals you do find, such as, water shrimps or caddis fly larvae, and build a project round them.

Survey as many streams as possible in this manner and produce accurate scale maps, drawings and bar graphs of the results.

Try keeping the species of flatworms in aquaria or model streams, at different temperatures and record your findings.

45   *(A)OE   Effect of altitude on the rate of growth of the moor rush* Juncus squarrosus

This is another summer project. The equipment required includes an Ordnance Survey map of the area (1:25,000), a penknife, ten or more polythene bags, ties and labels, and a 100 m measuring tape.

The moor rush is abundant on moors, bogs and moist heaths where the conditions are sufficiently acid. It grows at all altitudes. It forms dense low tufts with fibrous leaves from 8–15 cms long. The

Moor rush *Juncus squarrosus*.

flowering stems are 15–50 cms and very stiff and erect. The fruits are small brown capsules about 3 mm long.

Find an area where the rush is abundant at both higher and lower levels say 500–3,000 feet, but smaller ranges can be used. Locate your route on the map and the distance in metres between successive contour lines as in the previous project. Choose ten clumps of the rush at the highest level (randomly, if you think it necessary, see Appendix). Cut all the flowering stems as near as possible to the ground and enclose them in a bag with a note of the height. Measure the distance on the ground to the next 100 ft contour line and check the position with any land marks. Repeat collections at each 100 ft level.

Tabulate and graph the length of stem in cm, the number of flowers and the number of ripe capsules per stem against the altitude.

Repeat on other hillsides and in different areas if possible.

Try to formulate other questions relating to the growth and distribution of the plant and test your hypotheses in the field. Examples might be: a) How are the seeds dispersed?   b) How acid must the soil be?   c) Does the plant occur more in deer or sheep country?

46   A(O)   Distribution of the rush moth Coleophora caespititiella *in relation to the moor rush and to altitude*

This is an extension to the previous project to be carried out in late summer or autumn. September would probably be the best time.

The eggs of the moth are thought to be laid in June or July on or near the flowers of the rush. The larvae feed on the growing seeds inside the developing fruits and each forms a white cylindrical case on the fruit capsule by which its infestation may be told.

Plan a similar route to the one described above. Using a tally sheet, count the number of infected and non-infected fruit heads from the clump of plants that lies nearest to each 100 ft contour line. If possible count one hundred fruits. It may be necessary to devise a random way of sampling in order to avoid any choices that might lead to bias in the results (see Appendix).

Does the distribution of the moth bear any relationship to altitude?

Why do you think the moth is less widely distributed than the plant itself?

Can you devise ways of testing your hypothesis? See if you can find out more details of the life history of the moth at different seasons of the year.

47   A(OE)   Distribution of the common cotton-grass Eriophorum angustifolium *in relation to drainage and soil type (acidity)*

The equipment needed is an Ordnance Survey map (1 :25,000), a 100 m measuring tape, a metre rule, plant press, apparatus or materials for soil and water analysis.

The cotton-grass is very widespread in the North of England and the West Country, growing in wet bogs, shallow bog pools and acid fens. It is very easily recognised in June or July as the flower bristles elongate and become cottony, to which appearance it owes its name.

In order to avoid getting quite literally 'bogged down' it is suggested that this project is carried out from minor roads by walkers, cyclists, or car passengers with willing drivers.

With the help of the map and after extensive preliminary examination, choose a stretch of minor road from which patches of the plant can easily be observed. Map the distribution of the plant in relation to the road and for about 100 m from the edge of the road on each side. Choose several locations where the patches of cotton-grass can be safely studied. Devise ways of answering the following questions as accurately and quantitatively as possible.

Does the plant occur mainly in pure stands or mixed with other vegetation?

If the latter, identify the plants as far as possible. This may involve the use of the plant press (see Appendix).

What is the steepest gradient over 5 or 10 m from the edge of the area? How does this compare with gradients in non cotton-grass areas?

How does the water and soil from the cotton-grass areas compare with other areas in relation to pH and any other important factors?

What part may drainage and soil type play in the distribution of the plant?

48 AOE Lichen Projects

There is great scope here at any time of the year when the weather is favourable. These projects can be studied at any level and taken as far as time, interest and ability allow.

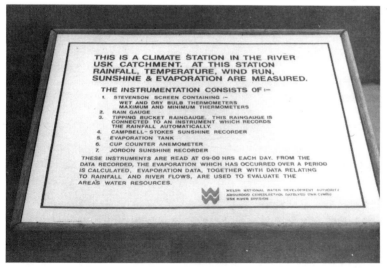

Climate station information. (Libanus Mountain Centre, Brecon Beacons)

48.1   Numbers of lichen species and their distribution on dry walls

This idea was prompted by the great number of lichen species on sandstone walls in mountainous districts of the Brecon Beacons National Park. It is wise to work within easy reach of a centre such as the Mountain Centre at Libanus near Brecon, especially in uncertain weather. Less time is wasted in finding a site for action and shelter; advice and information are close at hand, not to mention refreshments! There is also a weather station so that accurate records can be consulted of the conditions in which these plants flourish.

Problems of identification are considerable unless working with an expert but the *Oxford Book of Flowerless Plants* and the *Observers Book of Lichens* are useful.

Choose a wall with many different colours, shapes and textures of lichens. Each of these probably represents a different species. Spend a day collecting samples of each lichen with the help of a hammer and cold chisel. Collect enough of each to put one sample in a labelled envelope and another on a card with a protective transparent cover so that all the species can be viewed together. Until identification is certain, give each lichen a code number:

A conspicuous lichen (centre of picture) for detailed study.

White 1, 2 and 3; Grey 1, 2, and 3; Green 1, 2 and 3 and so on.

Select ten stations at equal distance apart along the wall and, using a transparent or wire quadrat frame (one decimetre square), record the number and code numbers of species occurring in a ten decimetre strip along the top of the wall and a similar strip on each side of it. How closely do the results from each of the ten stations agree? If there are any marked differences between the three surfaces, try to frame hypotheses to account for them and methods of testing them.

48.2 Study of one species in detail
Choose one of the distinctive species of lichens from the same wall and study its distribution in detail. Does it also occur on isolated rocks, on the ground or on trees in the same area?

48.3 Comparison of the distribution of crustaceous and foliose lichens
Crustaceous lichens are the most common forms on walls but foliose or leafy forms also occur. Compare the numbers of species and the area covered by each kind on walls at different altitudes, on different sides and the top of the same wall, on sandstone and

Leafy and crustaceous lichens on sandstone wall.

limestone walls, on tree trunks and walls in the same area.

48.4   Comparison of numbers of lichen species on limestone walls in different National Parks

Choose perhaps the Lake District and the Peak District. Can you relate any differences to pollution from industrial areas? Lichens are particularly sensitive to such pollution.

48.5   Open and closed lichen communities and competition between species

Open communities are those on which the whole walls or sections of wall are completely covered by lichens and there is no room for further growth. Look for such examples. Can you relate them to age of wall or rate of growth of different species of lichen? How many species are the closed communities composed of? Is there any evidence of encroachment of one lichen species on another? In such competition does the same species always win? Make drawings or take photographs to illustrate these features.

48.6   Adaptation of any of the above projects to isolated stones or rocks

48.7   Study of the lichens growing on gorse and heather

On gorse an adaptation of Project 61 can be used, using the diameter of the gorse stem and height from the ground in place of numbered internodes.

49   (O)E   Dead sheep

A somewhat grisly study that may appeal to certain students! The idea came from the observation of the number of dead sheep on main roads in the Brecon Beacons Park in Wales.

The peak death times are in spring when the young lambs are beginning to wander from their mother but will run back at the least sign of danger and in winter when the roads are salted and the sheep are attracted on to the road to lick. If these are not suitable times for observation, there may still be considerable numbers of casualties in the summer months as the heavier traffic takes its toll.

Work from the mountain side of the road, avoiding the traffic yourself. Locate several sheep which have been killed by traffic and

Sheep killed by traffic showing 'flies' and unravelled intestine.

are lying at the roadside. Enter the position on a large scale Ordnance Survey Map (1:25,000). Do not touch the carcase in case of disease. In each case try to estimate the numbers of hours or days since death and the extent of the injuries. Observe insect visitors and identify them if possible. They may be lured from the corpse and captured by baited pitfall traps (see p. 131) placed nearby. Observe bird visitors from a distance with binoculars and describe their activities. If you do not see the birds you may observe the results of their activities by seeing unravelled intestine or stomach nearby.

. If possible, keep the corpse under observation over a period of days or weeks and record developments. If it is removed try to find out who by.

You may be able to find a corpse which has been reduced to a skeleton. If so, remove the bones using old gloves. Disinfect the bones thoroughly and reconstruct the skeleton, naming the bones.

In spring or summer groups might stage a round the clock traffic census on roads where many, few or no corpses are found and try to find out whether more animals are killed at night and whether by cars or heavy vehicles. Also, watch to see how many sheep are seen on the road during the same period.

50   (A)OE   Effect of sheep or deer on moorland vegetation
Project 80 describes the building of cattle or sheep exclosures in a
field. This project takes advantage of exclosures constructed by
farmers or the Forestry Commission. These may be found in many
mountainous areas within the National Parks and elsewhere.

Find out by reading, personal enquiry and observation the effect
of sheep or deer on grassland and/or woodland vegetation. For
example, roe deer graze on grass and other herbs and in the process
destroy small seedling trees, preventing the natural regeneration of
forests; they browse on the shoots and buds of newly established
trees, crippling their growth and they fray bark with their antlers,
sometimes killing the trees back near to ground level.

Look for and enquire about areas where animals have been
deliberately excluded by the erection of wire netting fencing. The
most likely place may be where there is a young Forestry
Commission plantation. Find a place where there is public access to
the plantation or get permission to work in the area. By means of
random quadrats inside and outside the fence (see Appendix for
details) record differences in vegetation which should include the
height of plants, the number of species and the species composition.
You may be fortunate enough to find nearby areas where the
plantation has been enclosed for one, two, three or more years
making the comparisons more useful.

Smaller exclosures may include single trees or groups of trees
which have been fenced off for many years and where the difference
inside and outside the fencing is quite spectacular.

Take into account whether small animals such as rabbits can
enter the exclosure and the effects of manuring and trampling by
sheep and deer.

*51   A(O)   The plant and animal community of
heather-moor*
Find an extensive area of heather-moor in summer. It is best to
study one that has not been recently burnt. Burning is carried out at
intervals of 7–15 years on grouse moors.

Follow a path which runs roughly parallel to the contour lines.
Record the plant species found in 1 m quadrats, using frames
placed at 10 m intervals near to the path. To help you identify the
likely species and compare with the rest of Britain, here is a list of

plants from one hundred such quadrats, taken from an area ranging from Wales to Scotland. Most quadrats contained only a few of these species.

Flowering plants	Mosses
Calluna vulgaris, heather or ling	*Campylopus flexuosus*
Vaccinium myrtillus, bilberry	*Hylocomium screberi*
Deschampsia flexuosa, wavy hair-grass	*Hypnum cupressiforme*
Empetrum nigrum, crowberry	*Webera nutans*
Erica tetralix, cross-leaved heath	
Galium saxatile, heath bedstraw	**Lichens**
Agrostis tenuis, bent grass	*Cladonia coccifera*
Carex binervis, heath sedge	*C. floerkiana*
Erica cinerea, fine-leaved heath	*C. sylvatica*
Festuca ovina, sheep's fescue grass	*Lecanora conizeoides*
Juncus squarrosus, moor rush	
Nardus stricta, mat grass	
Potentilla erecta, tormentil	
Vaccinium vitis-idaea, cowberry	

METHODS OF STUDY

Various methods can be adopted to study the invertebrate (mostly insect) fauna associated with this plant community:

1 Hand searching or capture by pooter (see p. 84) of beetles, small moths, caterpillars, and spiders.
2 A strong wide-mouthed sweep net may be employed if found to be effective in the rather strong and prickly vegetation. Remove insects from the net by pooter or directly into the killing bottle (see Appendix).
3 Take a half metre square of stout, light coloured material and cut a slit half way along one side to the centre. Place this so that the slit encloses an individual heather stem or group of stems. Shake or beat the plants so that animals fall on to the sheet. Remove them quickly by pooter or put direct into killing bottle.
4 Leave water or pitfall traps out over night (see p. 131).

Insects of this community include the heather beetle *Lochmaea suturalis* which feeds on the bark and leaves, stripping the stem and sometimes causing plagues on damp grouse moors, leaving the plant a characteristic foxy red colour. There are a few other beetles and weevil species which feed on heather while predatory species include the striking green, crimson and copper *Carabus nitens.* There are fifty or more moth species whose larvae feed on heathers, heaths, bilberries and crowberries. Plant bugs, bees and spiders are also well represented and help with identification will be found in reference books (see Chapter 4 Nos. 5, 6 and 26).

Try to make the study of the fauna quantitative by counting the number of animals found in the quadrats used for the plants. Results will be dependent on weather, season and time of day, as well as the diligence of the search.

Further Reading

1   Chard, J. S. R. *The Roe Deer* Forest Record 99 H.M.S.O. 1974
2   Forestry Commission *Forest Fencing* Forest Record 80 H.M.S.O. 1972
3   Lyneborg, Lief (Ed. A. Darlington) *Dune and Moorland Life* Blandford (out of print)
4   Pearsall, W. H. *Mountains and Moorlands* Collins 1950
5   Reynoldson, T. B. *A Key to Freshwater Triclads* Freshwater Biological Association 1967

See also titles on the Further Reading Lists: Chapter 1 Nos. 4 and 16; Chapter 2 No. 15; Chapter 4 Nos. 1, 5, 6, 9, 18, 30; Chapter 5 Nos. 3 and 14.

4 Woodland

Even a small area of well managed woodland offers some possibilities for projects, but neglected woodland is much better, especially if there is plenty of dead and decaying wood. This nourishes many saprophytic fungi and scavenging animals and provides opportunities to study detritus food chains. Deciduous or mixed woodland is more rewarding than coniferous woodland.

It is usually important to obtain permission to work there from the owner of the woodland, unless there is general public access and the minimum disturbance can be guaranteed.

The Forest Commission Booklet No. 29, *Wildlife Conservation in Woodlands*, contains much valuable material which students will find helpful in some of these projects as well as being a source book for many others.

Woodland Projects

Projects based on trees

(A)OE 52   Growth
A(O) 53   Productivity of oak or other chosen species
(A)OE 54   Survival of tree seedlings
(A)OE 55   Effects of pollution on evergreen trees

Projects based on fungi

OE 56   Fungus forays in the autumn
AO 57   Bracket fungi (throughout the year)
A(O) 58   Fungi and insects
58.1   General study of relationships between fungi and insects
58.2   The insect associates of particular fungal species

Projects based on mosses and lichen

AO 59   *Leucobryum glaucum*
A 60   *Sphagnum* species
A(O) 61   Terminal twig invasion by lichens
A(O) 62   Vertical distribution of lichens

Projects based on ferns

AO 63   Variation in the common polypody
AOE 64   Bracken

Projects based on insects

AOE 65   The insects and other invertebrates associated with leaf litter
AO 66   Defoliating insects
AOE 67   Wood Ants

Projects based on birds and mammals

OE 68   Birds' nests
(A)OE 69   Rookeries
(A)OE 70   Grey squirrels
(A)OE 71   Further mammal projects

Effects of man

AO(E) 72   Effect of burning
(A)OE 73   Effects of litter dumping
AOE 74   Effects of wood-cutting
AOE 75   Effects of erosion

Projects based on trees (Projects 52–55)

52 (A)OE Growth

Choose a species which can be easily recognised, for example, the common oak *Quercus robur*, the beech *Fagus sylvestris* or the silver birch *Betula verrucosa*.

If this is a group or form project it will be more interesting if several different species are compared, or different individual trees of the same species are chosen.

Record times of bud opening, flowering, leaf colour change and leaf fall. To gain the maximum amount of information and interest, record the first date when any bud shows signs of opening and the date when all buds are open.

Take photographs or make accurate drawings from life. Refer to books but do not copy from them.

Relate biological data to temperature, rainfall, hours of sunshine, altitude and any other relevant physical factors.

In the winter, measure the length of new growth from the top set of girdle scars (where the terminal bud scales fell off last spring) to the new terminal bud. Measure this at different heights on the tree choosing ten twigs at each height, and on the north, south, east and

Horsechestnut *Aesculus hippocastanum* bud opening in early May.

west of the tree. Show the information in tables and graphs and analyse it statistically to see if there are any significant differences between individuals of the same species and of different species.

In school projects the information can be passed on from year to year to build up a valuable reference source.

53   A(O)   *Productivity of oak or other chosen species*

Productivity is the amount of living material formed per unit area per unit time. It is difficult to estimate this accurately, especially at an elementary level, but such a project is worth attempting for the insight it gives into ecological energetics which is the flow of energy through an ecosystem.

This is a suitable project for the autumn term, involving a small group over a long period or a larger group, over a shorter period.

Before leaf-fall starts, choose a moderate sized tree with plenty of acorns (or other fruit if a different tree is chosen). Find one well isolated from other oak trees if possible. Estimate the number of main branches on the tree. Select one of average size and estimate the number of smaller branches. Count the number of leaves on an average sized smaller branch. By this method, or a more accurate one if you can devise it, estimate the number of leaves on the tree. Collect one hundred leaves. Dry the leaves in a low oven or incubator until there is no further change in weight. Calculate the weight of leaf material produced by the tree in one season. If desired, find the chemical energy or calorific value of the leaves by the use of a bomb calorimeter (see *Nuffield Biology* Book 3).

Pick up all the acorns dropped by the tree throughout the autumn. Include the acorn cups. Dry one hundred acorns and the cups by the method given for the leaves. Calculate the weight of fruit material produced in one season and find its calorific value.

Using an adaptation of the method described in Project 1 on growth calculate the weight of the new growth of stem during the year. There will also be new root growth, but it will prove more difficult or impossible to estimate this.

The amount of energy from the sunlight falling on one square metre of land in temperate latitudes is in the order of 2.5×10^8 cal/m²/yr. From 95–99% of this is lost from the plants in sensible heat and heat of evaporation. The remaining 1–5% is used in photosynthesis.

From these figures it may be possible to get a very rough estimate of the amount of incident energy that is transformed by the tree into the chemical energy locked up in the leaves, fruits and new stem growth.

N.B. This will be net production which is equal to gross production minus respiration. Also note that it will be lower than the actual value as some of the plant production, such as leaves and acorns, will have been eaten by herbivores.

While collecting the leaves and acorns, notice that some of them have been attacked by herbivores such as caterpillars, while some of the acorns are carried away by birds and mammals.

As an interesting extension of the project, find the proportion of the leaves and fruits eaten by herbivores and those which go to decomposers such as fungi and bacteria. Select ten leaves at random (see Appendix) and draw round their outlines on graph paper. Indicate the eaten portions and calculate the percentage of the total area eaten.

Compare the number of cups and acorns collected. This gives a clue to the activities of larger herbivores like squirrels or birds which may eat or carry off the nuts. Examine the acorns for exit holes of insects whose grubs consume the seed. This could well lead to yet another project on the fate of acorns produced over one season. How many are eaten? How many rot? How many germinate and how many grow into sizeable trees?

It is worth noting that the acorn crop can vary quite dramatically from year to year.

54 *(A)OE Survival of tree seedlings in their natural habitat and in the laboratory*

Choose an isolated tree (if possible) which is surrounded by young seedlings. Measure the distance from the parent tree of the nearest and most distant seedling. If possible, count the total number of seedlings. Choose several one metre or half metre squares. (Larger metric areas may be chosen if necessary.) Record the position of each tree in the quadrat (see Appendix) and also record its height and number of leaves. If possible, take a series of quadrats at increasing distances from the parent. Be sure to mark the position of each quadrat by some distinctive method. Repeat at monthly or bimonthly intervals and assess the effect of distance from the tree and other factors on the rate of growth and survival of the seedlings. Make a private and public graphical record so that future students can build on your results.

Try transplanting some of the tree seedlings and grow them in gardens or in the laboratory. Lift them on a wet day and take a good ball of woodland soil around the roots.

It has been suggested that we should be wise to transplant some of this super-abundance of young tree seedlings to suitable places in the countryside where they may replace some of those damaged by drought, fungi or insects. More than twenty million trees have been lost in the last twenty-five years. Check whether permission is required to carry out such transplanting.

55   (A)OE   *Effect of pollution on evergreen trees*

This is an adaptation of a method given in *Nuffield Biology*, Book 4.

Choose an area of woodland that is near a busy road and which contains some evergreen trees and bushes, such as pine, ivy, holly, laurel, rhododendron, and yew.

Select trees or shrubs of evergreen species that are growing near the road. Measure the distance from the tree to the road. Pick about ten leaves from the part of the stem which has grown in the current year. The bark will be absent or very thin and the rings or girdle scars mark where last year's terminal bud scales fell off. Pick ten or more leaves from the previous year's growth, below the first set of girdle scars and ten leaves from the year before last. Keep the three sets of leaves in three separate labelled polythene bags. Repeat at measured distances from the road using the same evergreen species and/or different species.

Pull each leaf through a piece of folded white paper using the same pressure in each case. (Some practice may be necessary to standardise the method.) In the case of large leaves, portions of the leaf may be used and it is wise to cut the prickles off holly leaves. Label each strip carefully and stick into book or on file paper very neatly under sellotape.

Does the evidence support the hypothesis that   a) the older the leaf, the dirtier it is?   b) the nearer the road, the dirtier it is? What do you think the dirt consists of? What harmful effect may it have on the plant?

Take a traffic census to see how many vehicles pass throughout the day. Compare the results with a woodland area near a less busy road.

Projects based on fungi (Projects 56–58)

56 OE *Fungus forays in the autumn*

Preliminary work must be carried out to determine whether the area is suitable for such a project and it will prove far more worthwhile if a knowledgeable person is available to help. If the teacher is not an expert in this field, a local natural history society or museum may be able to suggest someone to consult or give advice themselves. Failing this, it should be possible to identify a number of the plants collected with the help of books. This tends to be time consuming and most of the fungi are short lived and only remain typical and beautiful for a few hours.

One advantage of studying fungi is that as they are only the fruiting bodies of the plant, collecting them need not be destructive. The mycelium or plant body remains in the ground or wood to produce further fruiting bodies.

Poisonous species are relatively few in number and easily recognisable, but it is wise not to eat any of the fungi collected and to wash hands carefully after the foray.

Equipment should include baskets for collecting. Chip or fruit baskets or any old wide based baskets are suitable. Never collect fungi in polythene bags, but reserve these for enclosing separate robust specimens to be kept apart from other more delicate ones. A trowel is needed to dig up mushrooms or toadstools as the base of the plant is often a diagnostic feature. A penknife is needed to cut off bracket fungi from trees or stumps.

Although it does no harm to collect a few fruiting bodies of fungi, moderation should be observed, taking only those really needed for study and leaving plenty so that others may benefit from seeing these very attractive plants whose season is usually very short. After a rainy spell in September or October is usually the best time to find the greatest variety of species and numbers of plants.

All fungi should be handled gently as they are easily broken and bruised. Place them carefully in the basket, not piled on top of one another. Keep separate containers for sub-habitats within the wood such as collections made under different species of trees or in clearings; these are often diagnostic for identification.

As well as collecting the more obvious mushrooms and toadstools, look for bracket fungi on dead tree branches or stumps, jelly-like and 'burnt' forms on wood and tiny delicate forms in a variety of habitats. Smell out the stinkhorn which develops from a subterranean jelly-like egg into the stage which gives its name of *Phallus impudicus*. A preliminary browse through some good illustrated books gives an idea of the range of forms.

On return to base, remove all the fungi carefully and place them in suitable containers, grouping those from similar habitats together. Record the collection by photography in the field and laboratory, quick sketches, painting and detailed notes. There is no satisfactory way of preserving fungi, although they can be placed in separate polythene bags and put in the deep freeze. They will deteriorate fairly quickly on removal. Woody bracket fungi keep longer.

Make gill prints by cutting off the stem portion (or stipe) from a fresh, fully expanded specimen and place the cap, gills downwards on a piece of dark paper. Use white paper for species with dark spores. Place a beaker or jam jar over the cap to prevent movement

Bracket fungus *Ganoderma applanatum* on tree stump.

of cap or spores. Remove after several hours (overnight). Make the print permanent by spraying with pastel fixative from the art room.

Bracket fungus growing on dead birch stump showing sawdust-like frass formed by fungus beetles. Note also epiphytic green alga *Pleurococcus* growing on upper surface of fungus brackets.

Bracket fungus and lichen on alder trunk.

57 AO   Bracket fungi throughout the year

Most mushrooms and toadstools are short lived fruiting bodies, but both annual and perennial bracket fungi are more long lasting and may be studied throughout the year. They grow on tree stumps or (usually) dead branches of deciduous trees. Common ones include the birch bracket *Piptoporus betulinus*, the many zoned polypore *Coriolus versicolor*, the beef steak fungus *Fistulina hepatica* and the common ganoderm *Ganoderma applanatum*.

Choose to study bracket fungi in general or one particular species that is common in your neighbourhood. Identify the trees on which the fungus grows, whether on stumps, dead or living branches or both. Map the distribution of the affected trees and fungi. Mark chosen fungi with small blobs of waterproof paint and study the rate of growth and decay. Find out if the fungus has any economic importance (harmful to timber) and whether it is a parasite or saprophyte.

Cut thin sections of the bracket with a sharp razor blade and examine the cut edge under the low power of the microscope. Identify parts and make drawings and permanent slides using Hoyer's medium (see Appendix). Find out from books the season of spore production. Examine the spores and make experiments to grow them on a petri dish of sterilised nutrient agar. See methods of note making and drawing in R. Watling *Identification of Larger Fungi*. This book contains much useful information and suggests other projects.

58 A(O)   Fungi and insects

While collecting fungi it is quite likely that the insects which feed on them and breed in them will be noticed. Most people have seen maggots in wild mushrooms. The relationships between fungi and insects is well worth studying and if the students are careful and enterprising new observations can easily be made. Some workers have studied such species as the birch bracket, the stinkhorn and *Daldinia concentrica*, while the boletus type fungi and many of the bracket fungi are also known to yield interesting results.

58.1   General study of relationships between fungi and insects

Start this in the autumn, examining the toadstool and mushroom type fungi for the presence of insect eggs, larvae, pupae and adults.

Boletus species showing part eaten by herbivores, probably slugs.

The *Boletus* species which have pores in place of gills are often colonised. The remains of the specimens from the fungus foray can be used. Enclose these individually under beakers (block the exit), or bell jars for a few days. Larvae may be seen and some beetles or adult flies may emerge. Study these under the stereoscopic microscope or hand lens. Draw and describe them.

Generally speaking, dipteran (fly) larvae have no legs, beetle larvae have three pairs of legs and lepidopteran or moth larvae have three pairs of true or thoracic legs and usually four pairs of abdominal or false legs. Insect larvae and pupae may be identified with the help of H. F. Chu *How to Know the Immature Insects*. It will probably not be possible to identify the species precisely, but much interesting work can be done, given curiosity and patience. See the Appendix for the use of Hoyer's Medium for making permanent slides which can be used to draw from later.

Collect other cap type and bracket fungi throughout the year. Look for those which are slightly mushy or which show sawdust like frass where insects have been feeding.

Classify the fungi into: a) those which never yield insect stages, b) those which sometimes do, c) those which often do and, d) those which always do.

58.2 *The insect associates of particular fungal species*

Recommended fungi are the birch bracket, *Daldinia concentrica* and *Coriolus (Trametes) versicolor*.

Birch bracket fungus *Piptoporus betulinus*.

The birch bracket is a common parasite of birch trees which kills them and then lives on as a saprophyte. Paviour-Smith has shown that the beetle. *Tetratoma fungorum* attacks the living fungus, while another beetle, *Cis bilamellatus* helps to destroy the old fungus. It would be interesting to see whether students can find and identify these beetles and develop the project according to their interests and abilities. (See Charles Elton *The Pattern of Animal Communities*.)

The black ascomycete fungus *Daldinia concentrica* grows mainly on dead ash, but is sometimes found on beech, oak and burnt birch and gorse. It is widely distributed and common. When fully grown

The Ascomycate fungus *Daldinia concentrica* on dead ash. Newly formed fruiting bodies below right; old eroded fungus above, showing centipede inside empty crust.

in September the larger fruiting bodies are dark brown and the size of half an apple. Before spore discharge the following May they are very dense. Inside, the concentric rings of tissue provide its specific name. After spore discharge the fruiting body becomes much less dense and its crust is black and often pitted with holes.

This fungus has three or more specific beetles, one specific fly and one specific moth associated with it. Specific insects breed only in this fungus and nowhere else. Many other insects and micro-animals have also been found inside the fungus.

Identify the fungus and note which trees it is growing on. Make distribution maps. Examine 'new' and 'old' fungi for signs of insect activity such as holes, eroded portions and frass (undigested remains of food). Keep some intact fungi on damp blotting paper in jars with muslin or nylon tight tops, secured by elastic bands. These jars can be kept for months; the blotting paper should be kept slightly moist but the fungus and paper should not be allowed to turn mouldy. Fungi collected in autumn should be kept at room temperature until the following May or June when the adult insects emerge.

Carefully cut up other fungi and look for the larvae and pupae of the specific beetle, *Biphyllus lunatus*. The beetle is easily recognised as it gets its name from the crescent shaped lighter patches on each of its elytra or wing cases. With care it is possible to rear the beetle through all its stages if the eggs, larvae or pupae are enclosed with a few pieces of sterilised fungus. (Heat some chopped up pieces for an hour in a fairly hot oven.) They should be placed in a petri dish with a circle of damp filter paper or blotting paper. Keep the paper slightly damp and view the dish under the low power of the microscope to follow development. Eggs occur in early spring, larvae from April to about August, pupae and beetles in the summer and autumn. The adult beetles overwinter in the fungus and may be found at all times of year. Be especially careful to record whether the insect stages are found in 'new', spore producing or 'old' fungi.

Coriolus, or as it used to be called *Trametes versicolor*, is much more common than either of the previous fungi. It is the attractively coloured, many zoned, thin bracket fungus growing mainly in layers on old tree stumps. While only a few *Daldinia* may be colonised by animals, *Coriolus* appears to have a more varied and abundant fauna.

Bracket fungus *Coriolus versicolor* on beech stump.

Study the fungus at various ages. Usually young and old stages occur together and can be easily distinguished. Collect about ten brackets of two or three different age groups, enclosing each group in a separate bag. On return to base, draw round each bracket on graph paper and note any visible signs of animal life on upper and lower surfaces, examining with a hand lens or stereoscopic microscope. Cut strips a few millimetres in thickness and examine the cut surfaces under the low power of the ordinary microscope, using top illumination. Draw and describe the animals found.

Follow the instructions for *Daldinia*, keeping intact brackets in jars and larvae with sterilised food in petri dishes. Keep detailed notes, drawings and observations, weekly over a period of months. Kill any adult beetles or other insects in concentrated ethanol and mount or preserve them as described in the Appendix. Identification can be attempted using K. Pavoir-Smith and J. B. Whittaker *A Key to the Major Groups of British Free-Living Terrestrial Invertebrates*. Precise identification may only be possible by experts and they are very busy people, but for O and A level work, Beetle 1, 2 and 3 and Fly A, B and C with drawings and descriptions is quite sufficient for the examiners, although perhaps frustrating for the researcher. Identification, although important, is not everything.

Projects based on mosses and lichens (Projects 59–62)
One advantage of these projects is that mosses and lichens are abundant and found throughout the year. They can be kept fresh for a long time and are easily preserved. They provide excellent winter projects. A disadvantage is that many of them are difficult to identify. The projects suggested deal with easily identified species.

59   AO   Leucobryum glaucum
This moss forms very typical rounded cushions, bright green to whitish in colour and up to 10 cm high in the centre. It is abundant locally in woodland on poorer leached soils. It is becoming rarer, so that on no account should whole cushions be removed from the habitat. If small portions are removed, return them unharmed.

Study the distribution of the moss in relation to soil type, amount of shade, dominant vegetation, slope of land etc. Find a place where there are many patches in a limited area. Using half metre or one

The moss *Leucobryum glaucum.*

metre quadrats map the position and extent of each cushion of moss (see Appendix). Carefully mark the position of each quadrat and repeat the observations at monthly intervals.

Remove one or two plants from the outer part of the cushion. Mount leaves and portions of stem in water and make drawings through the microscope. Permanent slides may be made using Hoyer's medium (see Appendix) but they will not retain their colour. Small portions can be preserved by drying in newspaper and mounting or keeping in an envelope in a dry place. The addition of water to the dried moss will restore it to its previous form.

Take small portions of cushions to examine for animal life at different seasons of the year. If the moss is wet, squeeze drops of water out and examine under the microscope. If the moss is dry, add clean rain-water and squeeze out. Larger animals may be driven out of the moss by placing a small sample in a plastic funnel, placed in a conical flask, containing a few cm^3 of strong ethanol. Place a lamp as close as possible to the moss sample. The light and heat will drive small animals like mites or springtails into the flask, where they will be killed and preserved for examination. If preservation is required the ethanol should be diluted to 70% after the animals are dead. Permanent mounts can be made as described in the Appendix.

The same method can be used for other species of mosses and with the same species at different times of year and the results compared.

60   A.   Sphagnum *species*

The bog mosses are fairly widely distributed and may be found in marsh and heathland as well as in some woodlands. The species within this genus are not easy to identify, but this is not essential.

Sphagnum species.

The *Sphagnum* leaf is an object of great beauty and scientific interest when viewed under the microscope. Like all moss leaves it is only one cell thick and when mounted in a drop of water under a cover slip, the water storage cells and the chlorophyll containing cells can easily be seen. Often other plants such as diatoms and desmids, and animals such as rotifers and amoebae can be seen on the surface of the leaf or inside the cells. In fact, the *Sphagnum* plant is a micro-habitat for an amazing assortment of microscopic plants and animals. Samples of moss from different habitats may be collected and their distribution studied as in the previous project. Do not collect unless moss is plentiful.

If there is no *Sphagnum* available locally, a small handful of the moss can be collected on a field trip or while on holiday. Kept in an open polythene bag with the top just folded over to prevent desiccation, it will remain fresh indefinitely and so will the animals it contains. The changed conditions may bring about some alteration in the numbers and activity of each species. Counteract a tendency to dry out by the addition of a little rainwater from time to time.

The easiest way to obtain the species of plants and animals associated with this moss is to squeeze water from it onto a microscope slide, as described in the previous project. Many of the desmids, diatoms and other algae, protozoa and rotifers are described in John Clegg *Observer's Book of Pond Life*. Others such as the gastrotrichs, insect larvae, mites, worms, crustaceans may be found in the *Nuffield Key to Organisms in Ponds and Water Troughs*.

Amongst the commonest forms of animal associated with *Sphagnum* are members of the Testacidae which are amoebae with cases formed of sand grains, diatom shells or scales. There may be hundreds of these in one drop of *Sphagnum* water. Examine under a coverslip under high power to see pseudopodia and to make drawings of the various types. Make permanent slides for later study by allowing the drop to dry and immediately adding a drop of Hoyer's Medium (see Appendix) and a cover slip. See Further Reading List book No. 9 at the end of this chapter for identification.

Other observations could include movement in amoebae, ciliates and flagellates, especially *Euglena*. The opportunities are endless for someone with curiosity, patience and a liking for microscope work.

61   A(O)   Terminal twig invasion by lichens
Many, though not all, trees have a rich twig flora of lichens. The initial invasion of ash *Fraxinus excelsior* twigs begins on the third to fifth internodes. These are easily calculated by counting the spaces between buds.

Carry out a preliminary survey to see which trees in your neighbourhood have the richest twig flora of lichens. If you live in a polluted area this may present difficulties as lichens are very sensitive to such pollution. (See next project.) Choose one or more species of trees to study and identify the lichens which grow on the twigs. Collect a small sample of each lichen and make an

Lichens indicating minimum air pollution on oak branches. (St Mawes, Cornwall)

identification sheet by sticking each one under sellotape and giving it a code number until identification is certain. With the help of *Observers Book of Lichens* it should be possible to identify at least the genus of the lichen. Make out a table similar to the one shown which sets out the frequency percentage of lichens invading nodes of *Fraxinus americana* in Minnesota. Sample size for internodes 1–10, 10 trees; for internodes 11 and 12, 8 trees; and for internode 15, 4 trees. (The species of lichens you find may be different.)

62   A(O)   Vertical distribution of lichens

Tree saplings are probably colonised by lichens initially through the same process as twig invasion. As the tree trunk and branches grow, the lichen communities form a more or less continuous cover over the bark. As the crown forms, the lower parts of the trunk become shaded and the environment along the trunk is radically altered. The lichens become sorted out along the trunk in response to their different environmental needs. A characteristic pattern of vertical distribution emerges.

Make out an identification sheet of the lichens found on the trunk and main branches of chosen trees as suggested in the previous project, sticking samples under sellotape and coding them until identification is certain. Work in pairs for safety reasons, using a

	Internode														
	1	2	3	4	5	6	7	8	9	10	11	12	13	14	15
Physcia stellaris	—	—	—	50	100	100	100	100	100	100	100	100	100	100	100
Xanthoria polycarpa	—	—	—	—	50	90	100	100	100	100	100	100	100	100	100
P. adscendens	—	—	—	—	—	10	40	50	80	90	100	100	100	100	100
P. orbicularis	—	—	—	—	—	10	20	40	70	90	100	100	100	100	100
P. ciliata	—	—	—	—	—	—	10	20	30	50	50	50	57	57	50
Lecidea melancheima	—	—	—	—	—	—	—	10	—	10	50	50	43	50	25
Caloplaca cerina	—	—	—	—	—	—	—	—	20	50	63	75	71	71	100
Lecanora subfusca	—	—	—	—	—	—	—	—	10	20	38	63	71	86	100
Candelariella vitellina	—	—	—	—	—	—	—	—	—	10	12	25	43	57	75
Rinodina pyrina	—	—	—	—	—	—	—	—	—	—	12	50	86	71	75
Candelaria fibrosa	—	—	—	—	—	—	—	—	—	—	25	12	14	—	25

Table showing % frequency of lichens invading nodes of *Fraxinus americana* in Minnesota.

Usnea ceratina growing on blackthorn.

Ash trunk showing epiphytic moss and lichens.

Elm trunk showing lichens and ivy.

Beech trunk showing several species of lichens.

ladder if necessary. Take light readings and samples at say two metre intervals. A wet and dry thermometer can also be used to find evaporation rate (see Appendix).

Carry out experiments to transplant colonies of lichens to different levels on the trees. Remove the lichens very carefully with the underlying bark and tape it firmly at a different level. Record growth or decay of colony over several months. Transport some colonies from upper levels to lower levels and vice versa. Even changes of one metre have been found to be important, but more work remains to be done in the way of original research.

Projects based on ferns (Projects 63, 64)

(63) *AO   Variation in the common polypody* Polypodium vulgare

This common fern is very easily recognised and grows throughout the British Isles in hedgerows and banks as well as in woods. It has lobed leaves which vary in quantity, length, width and in number of lobes as well as in sori (reproductive organs) per leaf. These variations may be genetic or environmental.

Choose an area where this fern is abundant. Choose ten or more plants which are fairly close together and for which the environment seems similar. Monitor the environmental conditions by taking temperature, light and humidity readings beside each plant. Fill in the following table with the relevant information.

Temp-erature	Light	Humidity	No. of leaves	Average length of leaf	Average width of leaf	Average no. of lobes/leaf	Average no. of sori/lob
1							
2							

Show the postion of the plants on an accurate scale map.

Repeat using ten plants from different environmental conditions, for example, damp, dry, very shady, less shady, epiphytic on tree, isolated, or near other ferns and so on. Make out a similar table and map. (For quantitative work collect more data.)

Make bar graphs and analyse statistically to test whether environment affects variation in this fern.

Extend the study to other areas visited on expeditions and on holiday.

There is no need to take any plant samples as all the measurements can be made in situ.

64   *AOE   Bracken* Pteridium aquilinum

This plant has been chosen because it is so common and easily recognised and because it grows very quickly and has other interesting features.

Begin this study at any time from May, when the young fronds are appearing, until late autumn when they are dying down, although it is best to locate the area to be studied in summer when a suitable patch can be identified most easily.

Choose one or two contrasting areas of limited size, say five metres square, one in the open and one under trees. When the plants are fully grown, record the percentage ground cover in each area using half metre quadrat squares (see Appendix). Using a measuring tape as a base line, peg out the corners of each square, leaving these pegs as permanent markers. These should be well hidden but easily recognisable. Fasten a length of nylon cord or string between the pegs to make a square. With the help of the measuring tape and a metre rule and squared paper mark A–J along the top and 1–10 along the sides, record the percentage cover by bracken in each of the one hundred squares. This is most easily done by shading in the approximate area covered by bracken in each square. Take five squares at random and record the vegetation in more detail (see Appendix for use of quadrat and for random sampling).

Take light readings with a light meter and/or probe at monthly intervals throughout the season. Analyse soil samples from both areas for moisture and humus content. Relate these and any other relevant physical factors to the differences in percentage cover by bracken in the two areas.

Record the dates of emergence of the first new fronds in early summer, probably May, choose ten plants at random (see Appendix) from each area each week. Record the rate of growth of the fronds in tables and bar graphs. Relate to weather conditions (temperature and rainfall). Take photographs and/or make drawings to show the development of the plant throughout the season. Record the development of the sori on the under surface of the leaf segments. Look for ripe spores during July and August. Try germinating these on damp filter paper to which a little of the soil in which the bracken is growing has been added. Use a petri dish and lid. The spores develop first into small prothalli which bear the sex organs. (See a botany textbook for details of life history.)

When the leaves have almost died down, dig up a whole rhizome or underground stem. This is easier to do at the side of a path where the soil is cut or washed away. Examine the rhizome system of the plant. Make drawings to show the old and new sections. A new section is formed each year from the food sent down by the dying foliage leaves. Test the rhizome for starch and other food substances. Find the fresh and dry weight of a cubic centimetre of the old and new sections of the rhizome to show whether food reserves have been used up during growth.

The productivity of a bracken plant during a season can be measured using a modification of the method given on p. 60. The bracken is a perennial plant and the rhizome remains below the ground during winter. The plant is able to make very rapid growth in summer, using the stored food in the rhizome.

There is a theory that bracken produces poisonous chemicals which prevent competition from other plants. Investigate this by collecting soil amongst which bracken grows and samples of other woodland soil. Try growing a selection of woodland plants on both soils under controlled conditions. Use healthy seedlings and plant them in seed trays enclosed in polythene bags until the seedlings are established. Keep careful records over a period of time.

Search carefully for any evidence of bracken being eaten by animals (including caterpillars and so on). If your results are negative suggest possible reasons and plan further investigations.

Projects based on insects (Projects 65–67)

65   AOE   The insects and other invertebrates associated with leaf litter

Collect samples of fallen leaves in late autumn, winter, spring and summer. In each case collect from the same area in a standard sized polythene bag. Weigh the bag plus litter immediately on return to base. Empty the bags, one at a time, on to large trays or developing dishes. Have a number of stoppered tubes or petri dishes plus lids available and capture any large animals, such as spiders and beetles, in these. Examine and describe quickly or kill and preserve for identification later (see Appendix).

Place the rest of the sample in a large plastic funnel and collect the smaller animals as described on p. 73. Count the numbers of each kind of animal and make permanent slides of each representative type as explained in the Appendix. Draw and identify as far as possible.

When each litter sample is completely dry, reweigh and calculate the percentage of water present. It may be worth while to collect after dry and wet spells at each season and find out whether there are any differences between the numbers and species of the animals collected.

This project can be extended in various ways such as collecting at different depths and under different trees.

66 AO Defoliating insects

Choose oak, beech, hazel or other suitable tree or shrub. Start the project in May as soon as the leaves are open.

Select one of the lower branches at random, using the following method, or devise a more suitable one. Using a small field compass stand on the north side of the tree. Draw a plan of the tree, marking compass points N, S, W and E. Mark off the number of easily reachable branches in each section and number them 1–10 plus (see diagram in Appendix).

Leaves of lime *Tilia europaea* showing insect damage.

Take the same number of strips of paper as there are branches and number accordingly. Get a friend to select a piece at random (draw lots). When branch is selected, carefully remove ten or more leaves (multiple of ten). The number will depend on how much time is available or the number working in the team. If you consider it is necessary, devise a way of selecting these leaves at random (see Appendix). The purpose is to avoid any subjective choice, such as taking those which are more eaten than others.

If it is safe and possible to climb the trees, the above procedure can be repeated at two metre intervals of height. Place leaves from different heights in different labelled polythene bags.

While collecting look for caterpillars or other possible defoliating insects and put these in a separate bag with a sprig of leaves.

On return to base place any live insects on the sprig of leaves in a small container of water with the water surface blocked with cotton wool or covered by plastic so that the insects do not fall in. Cover with a large polythene bag with air holes or use a proper insect cage. This consists of circular tin with small holes in the lid, the tin and lid being separated by a large thick cylinder of polythene. Rear the caterpillars in this container, supplying them with fresh food as required. Put slightly damp leaf mould and soil in the base of the tin as they may require this for pupation. When the adults emerge they are easier to identify than in the larval stage.

Draw round each leaf on graph paper. Include any eaten portions. Calculate the percentage of leaves eaten.

Repeat the whole procedure at monthly intervals until leaf fall.

In June and July, or later, spread a white sheet below the tree and shake or beat the lower branches to dislodge any insects. Collect these insects quickly using plastic petri dishes with lids, or use a pooter (consisting of a small plastic or glass tube provided with a two holed bung). The bung contains one bent glass tube and a straight tube with a rubber mouth piece attached; the other end of the straight tube is covered by gauze. By placing the open end of the bent tube over the insect and sucking at the mouthpiece, the insect is sucked into the bottle.

On return to base, identify the insects as far as possible and decide which may be responsible for eating the leaves. If some of these are caterpillars, rear them as described above.

Wood ant nest.

67 AOE Wood Ants

Before starting such a project, consult and read at least one of the books on the further reading list at the end of this Chapter, especially Nos. 4, 44 and 45 to gain an insight into the fascinating life of these insects.

Old colonies may be as much as five or six feet high and ten or twelve feet in diameter and can easily be recognised, consisting of mounds of twigs, old bracken and pine needles.

Since the depredations of aquarists to obtain pupae (sold as ants' eggs), this insect has almost become an endangered species. If it is common locally a magazine like *The Amateur Entomologist* would probably be glad to receive accurate details of the number and position of colonies in your area. Map them on a large scale Ordnance Survey map and make notes on them. This is a useful task in the autumn while some activity can still be observed.

Examine the colony on warm days in early spring to observe the first activities. Spend about an hour each week or month observing the movements and habits of the workers, noting what each is doing or carrying and how far afield from the nest they go. Track down their aphid farms on nearby trees. Measure the distance travelled by the ants on their foraging expeditions and plot this against time of year. When food is scarce in early spring they travel much further than later in the year.

Interfere with the nest as little as possible for conservation and safety reasons. (There is a danger from formic acid.) If you can obtain a few winged queens and workers when they are active on the surface, keep them with some of the materials they use for nest building, in a suitable container at home or school to make further observations. (See Reading List for further details.)

Projects based on birds and mammals (Projects 68–71)

The majority of children are naturally more interested in larger animals than they are in plants and insects, but these are much more difficult to study successfully owing to their rapid movements and secretive habits. For this reason only a small number of projects are included in this section, but the references include books with plenty of ideas for those whose interests lie in this direction.

68   OE   Birds' nests

This is a good winter project which requires no equipment except a few polythene bags and keen powers of observation and patience to search.

Examine bushes and trees after leaf-fall to find empty nests. Collect as many as possible, placing each in a separate polythene bag with a note inside giving the location and exact site.

On return, identify each nest with the help of Collin's *Guide to Nests and Eggs*. There may be abandoned eggs inside which aid identification. Describe, sketch and photograph each nest. Take each nest systematically to pieces, collecting any insects which emerge and preserving them for later identification, drawing and description. Divide the materials used in the nest construction into several piles according to whether they consist of twigs, grass, moss, or other materials. When all materials are quite dry weigh them to find out the percentage by weight of each component. Stick a sample of each component neatly under clear adhesive tape for your record.

Find out from books, and from observations next spring, the way in which the different species of birds construct their nests.

69   (A)OE   Rookeries

This is another good project for winter and early spring and more interesting if it covers a considerable area. Bicycle trips are recommended, but walkers and weekend car travellers can make useful observations too.

February is a suitable time to start as then the communal nesting sites can be identified by the raucous noises of the rooks rebuilding, as well as by spotting the nests at the top of tall trees. The nesting sites may be in woodland but more often in clumps of tall trees amongst agricultural land. Most favoured trees are elm and oak. Owing to the destruction of many elms in recent years by Dutch Elm disease, rook-nesting habits are undergoing adaptations which should make an interesting study.

Locate as many rookeries as you can in your neighbourhood and in the surrounding countryside during weekends and holidays. Mark the positions accurately on a large scale Ordnance Survey map, or maps. Make tracings of these are required. For each rookery fill in a table as follows:

No.	Grid Reference	No. of nests	No. of trees with nests	Average height of nests	Species of trees

Make recordings of sounds at the rookery if you have a portable tape recorder or can borrow one. Record in different months and at different times of day. Try to relate sounds to activities.

Study one rookery near home in detail, noting your observations. Without risking life and limb try to get hold of an old nest to examine as described on p. 86. If a dead rook is available make drawings to show adaptations to feeding. Dissect to find contents of crop and relate to observations on feeding habits.

70 (A)OE   Grey Squirrels

Those who are fortunate enough to live in those parts of the country, mainly in the West Country and the North of England, where our own native red squirrel is still abundant, can follow the same pattern of study as outlined for the introduced American species *Sciurus cariolensis*. Grey squirrels often have a touch of red around the head but lack the characteristic tufted ears and very bushy tail of the native species.

It may be useful to start by clearing up two common misconceptions. Firstly the grey squirrel has not actively 'driven out' the red species by fighting it or even eating it as some children think. Certainly the grey squirrel has reached pest proportions and is still extending its range, but the exact nature of its competition (if any) with the red species is not fully understood. Secondly, neither species of squirrel hibernates during the winter. While they may spend considerable periods sleeping in their dreys during cold spells both species are active on sunny days throughout the winter.

For young students, it is often easier to study the activities of squirrels at second hand, as much patience and time is needed for squirrel watching. Locate their dreys which are like large untidy nests high up in trees, often in oaks. Look for evidence of activity near the dreys. This includes the characteristic stripping of pine cones, leaving only a rough kind of stem with all the scales bitten

Bark stripped by grey squirrels.

right off. They eat the winged seeds inside. Nuts eaten by squirrels are split neatly in two halves. Often tree stumps can be found on which the squirrel has deposited the pine scales together with the nut shells and a pile of its droppings. Collect this evidence, putting that from each stump into a separate bag, listing and describing the contents on return to base. After clearing each stump number it with waterproof paint and return at weekly intervals to see whether squirrels have been active on the same stumps. Collect and record this evidence.

Grey squirrels are notorious for the damage they do by stripping bark from trunk and branches, especially of young trees, and for uprooting bulbs. Look for and record this evidence with notes, drawings and photographs. Also record any evidence you may see of squirrels taking eggs or nestlings of birds.

71 (A)OE Further mammal projects

A useful book which gives further ideas for the study of mammals and other references is G. B. Corbet *Finding and Identifying Mammals in Britain*. There is a good section on trapping small mammals and a description of the part which amateurs can play in local and national recording. Other books in the Reading List for this chapter give valuable information on the study of tracks, trails and signs and the use of owl pellet analysis as an indicator of small mammal distribution.

Effects of man (Projects 72–75)

There is hardly any of the original natural woodland left in Britain and none at all that has not been greatly influenced by man. Possible projects on this subject are endless, so a few relatively simple ones, likely to yield interesting results in the short term, have been chosen.

72 AO(E) Effects of burning

Deliberate and accidental fires are all too common in woodland. Where woodcutters have been at work there are often small areas where they have burnt the brushwood. Find such an area soon after burning. It is useful to know the actual date of the fire. Using the quadrat method and an area not more than two metres square, keep a record over a period of months of the vegetation which colonises the burnt patch. If you cannot recognise the plants when they are young, take a similar sample from outside your study area and press, mount and draw it to help with later identification. Some flowering plants, for example rosebay willowherb *Chamaenerion angustifolium* and mosses, such as *Funaria hygrometrica* and fungi, such as *Peziza* species are frequently found in such situations. Watling in his book on *Identification of the Larger Fungi* gives a list of twenty-five fungal species that grow on burnt ground.

Peziza species of cup fungus growing on recently burnt ground.

At the end of your study, which should occupy at least a year, make a detailed plan of your quadrat and its vegetation on squared paper and another of a nearby area which was not burnt for comparison (see Appendix).

73   (A)OE   Effects of litter dumping

The quantity of assorted litter and junk dumped by the general public, especially on the outskirts and entrances to woods, is appalling. As well as having a direct impact on the fauna and flora it also affects the amenity value of the woods for recreational purposes.

Choose an area of woodland to which the public has free access and identify the main entrances, marking them on a tracing of the relevant part of a large scale Ordnance Survey map. Note entrances which have local Council warnings about the dumping of litter and any penalties that may be incurred. Classify each entrance and roadside area of the wood on a point scale, for example, 0:No litter, 1:Some litter, 2:More litter—mainly small items, 3:a great deal of litter, including large items such as mattresses, old cars and so on. Take photographs and measure the area covered by the litter. Assess the effect of vegetation. Is there more vegetation, or less vegetation, or is the vegetation different from similar nearby areas that are free from litter?

Visit the area at different seasons of the year, on week days and after the weekend to report on any changes. Contact your local Council to see if they are doing anything to clear up the litter. Clear a bad area yourself or with the help of the Council. See how long it takes to deteriorate. Publish your results in the form of a letter or article in the local paper and enlist local support in a campaign to clear up.

It is not very easy to assess the effect of litter on animal life, but dumping litter may not only drive away species, but also create new niches for a variety of animal life. Turn over stones, bottles, logs, or boards and record any invertebrate life such as wood lice, snails, centipedes and so on that may be found beneath them.

BEWARE OF BROKEN GLASS!

Wash out contents of old bottles and examine for skeletons of small mammals which have died in them. Identify species, mount skeletal remains on cards. Use data for distribution of the various species.

74 AOE *Effects of wood-cutting*

Find a place where trees have been recently felled and choose an area of say five metres square, and another similar area nearby where the trees have not been felled. These areas can be compared as described on p. 81 in the bracken study. Make a special study of a quadrat, say one metre square, that contains a tree stump. Compare it over a period of time with a similar area that contains an unfelled tree of the same species. Compare the physical factors such as temperature, light and humidity in the two areas.

As well as studying any changes in vegetation around the tree stumps, the stump itself may be investigated to see how it becomes colonised by mosses, lichens, fungi and flowering plants. To avoid

Cup fungus *Peziza repanda* growing on sawdust.

too long a wait for development, identify other stumps of less recently felled trees and those which have been long felled and compare the state of colonisation.

75   *AOE*   Effects of erosion

Woodland areas show few signs of erosion compared with arable land as the trees and other woodland vegetation help to bind the soil and prevent it washing away. This effect can be well appreciated if erosion is compared on paths and the woodland area through which they run.

Choose two gently sloping areas, one on a well-defined path and one amongst typical woodland vegetation near the path. Devise a way of measuring the angle of slope. Cut a small trench at the base of each slope to hold about a metre length of plastic guttering. Arrange the guttering so that it will collect the water from a two gallon plastic bucket poured on to the ground from a point a few metres above the guttering. If possible, carry out this project where water is readily available from a nearby pond or stream. Measure the amount of water which collects in each gutter (which must be sealed at each end) after a given time. Repeat on different slopes in different parts of the wood.

Further Reading

1   Alvin, K. L. and Kershaw, K. A. *The Observer's Handbook of Lichens* Warne 1963
2   Angel, Heather *Photographing Nature: Fungi* Fountain Press 1975
3   Preben, B. and D. *Collin's Guide to Animal Tracks and Signs* Collins 1972
4   Brian, M. V. *Ants* Collins New Naturalist No. 59. (1977)
5   Burton, J. and Yarrow, I. *The Oxford Book of Insects* O.U.P. 1973
6   Chinery, M. *A Field Guide to the Insects of Britain and Northern Europe* Collins 1973
7   Chu, H. F. *How to know the Immature Insects* Brown 1949
8   Corbet, G. B. *Finding and Identifying Mammals in Britain* British Museum of Natural History 1975
9   Corbet, S. A. 'An illustrated introduction to the testate Rhizopods in Sphagnum with special reference to the area around Malham Tarn, Yorkshire' *Field Studies* 1973, 3; 801–838
10   Darlington, A. (ed.) *Woodland Life* Blandford 1972
11   Duddington, C. J. *Beginners' Guide to the Fungi* Pelham Books 1972
12   Edlin, H. L. *Trees, Woods and Man* Collins New Naturalist 1970
13   Edlin, H. L. *A Guide to Tree Planting and Cultivation* Collins 1975
14   Elton, Charles *The Pattern of Animal Communities* Methuen 1966
15   Findlay, W. P. K. *Wayside and Woodland Fungi* Warne 1967
16   Forestry Commission *Know Your Conifers* H.M.S.O.
17   Goodfellow, Peter *Projects with Birds* David and Charles 1973
18   Hale, Mason E. Jr. *The Biology of Lichens* Edward Arnold 1967
19   Harrison, C. *A Field Guide to the Nests, Eggs and Nestlings of British and European Birds* Collins 1975
20   Hvass, E. and H. *Mushrooms and Toadstools in Colour* Blandford 1961
21   Lange, M. and Hora, F. B. *Collin's Guide to Mushrooms and Toadstools* Collins 1963
22   Lawrence, M. J. and Brown, R. W. *Mammals of Britain, their tracks, trails and signs* Blandford 1967
23   Matthews, Harrison *British Mammals* Collins 1952
24   Mitchell, A. *A Field Guide to the Trees of Britain and Northern Europe* Collins 1974
25   Neal, Ernest *Badgers* Blandford 1977
26   Nichols, D. and Cooke, J. *The Oxford Book of Invertebrates* O.U.P. 1971
27   Nicholson, B. and Clapham, A. *The Oxford Book of Trees* O.U.P. 1975
28   Nuffield Foundation *A Key to Small Organisms in Soil, Litter and*

Water Troughs Longman 1966

29   Nuffield IV *Living Things in Action* Longman 1966
30   Oldroyd, H. *Collecting, Preserving and Studying Insects* Hutchinson 1970
31   Paviour-Smith, K. and Whittaker, J. B. *A Key to the Major Groups of British Free-Living Terrestrial Invertebrates* Blackwell Scientific Publications 1968
32   Phillipson, J. *Ecological Energetics* Arnold 1966
33   Pursey, H. *The Wonderful World of Mushrooms and Other Fungi* Hamlyn 1977
34   Ramsbottom, J. *Mushrooms and Toadstools* Collins 1953
35   Randell, R. *Trees in Britain* Broadleaved Books 1, 2 and 3 Conifers and Allies Jarrold Colour Publications
36   Rinaldi, A. and Tyndale, V. *Mushrooms and Other Fungi* Hamlyn 1972
37   Simms, E. *Woodland Birds* Collins 1971
38   Steele, R. C. *Wildlife Conservation in Woodlands* Forestry Commission Booklet 1975
39   Stockoe, W. J. *The Observer's Book of Trees and Shrubs* Warne 1937
40   Stockoe, W. J. *The Observer's Handbook of Ferns* 1950
41   Sudd, J. H. *Introduction to the Behaviour of Ants* Arnold 1967
42   Watch *Tree Project Dossier* Advisory Centre for Education, Cambridge
43   Watling, Roy *Identification of the Larger Fungi* Hulton Educational Publications 1973
44   Wilson, Edward *The Insect Societies* Harvard University 1971
45   Wragge Morley, Derek *The Ant World* Penguin Books 1953
46   Yalden, D. W. *The Identification of remains in owl pellets* The Mammal Society

See also titles on Further Reading Lists: Chapter 1 Nos. 4, 6, 16 and 17; Chapter 2 Nos. 10 and 12; Chapter 5 Nos. 3 and 9.

5 Field, Hedgerow and Wasteland

The word *field* covers a wide variety of habitats including playing fields, cornfields and pastures. In the case of the last two it is important to observe the country code and to obtain permission if extensive work is intended. Be careful not to be involved in anything which might be construed as trespass.

Hedgerows are fast disappearing as fields become larger. They form important refuges of wild life and are still abundant enough to provide considerable opportunity for valuable observations and discoveries.

Wasteland is any uncultivated part of the countryside which is not woodland, heathland, moorland, mountain or marshland. It may also be a neglected plot in the suburb or town.

Field, Hedgerow and Wasteland Projects

Detailed autecological projects

(A)OE 76 The bramble *Rubus fructicosus* and the associated animals

(A)OE 77 The stinging nettle *Urtica dioica* and the associated animals

(A)OE 78 The wild arum *Arum maculatum* and the associated animals

OE 79 A study of the yellow hill ant

Field Projects

A(O) 80 The use of exclosures in a field to compare growth of vegetation inside and out

A 81 Grasshopper population study

A(O) 82 Colonisation of cow pats

AO(E) 83 Plant galls and their animal inhabitants

(A)OE 84 Variation in the number of floral parts in some common plants

84.1 Petal number in the lesser celandine

84.2 Stamen and carpel number in the lesser celandine

84.3 Stamen and carpel number in three species of buttercup

84.4 Floret number in the daisy or the dandelion

84.5 Number of flowers in the inflorescence of members of the Umbelliferae or Labiatae family

OE 85 Fruits, seeds and dispersal

Hedgerow projects

OE 86 Composition of hedgerow

86.1 Age of hedgerow related to number of plant species it contains

86.2 Birds, and insects of the hedgerow

86.3 Dangerous litter

(A)OE 87 Adaptation of plants to their position in the hedge and climbing plants

OE 88 Census of common birds

Detailed autecological projects (Projects 76–79)

76 (A)OE The bramble Rubus fructicosus *agg. and the associated animals*
The bramble occurs in fields, hedges and wasteland. It is widespread, common and easily identified. It is most variable with three hundred or more common species and it is used by many animals for food, shelter or breeding.

The two important reference books for this project are the *AA Book of the Countryside* and the *Flora of the British Isles.*

Use the checklist provided for making notes on each bush or clump in the field, filling in their positions on an accurate sketch map.

Specimens should be taken using a strong knife or secateurs. A separate labelled polythene bag is needed for each bush. Include a) a piece of stem from the middle of the section which is one year old, b) inflorescence (flower spray), preferably at two stages showing i) buds and flowers, ii) flowers and young fruit. Use these specimens to complete your checklist, for drawing and for possible identification of species with the help of the relevant pages in the *Flora.*

In autumn collect fruits, noting the times of ripening in the different varieties. Take ten at random (see Appendix) from each different type of bush and weigh them. Find the mean weight. Count the number of drupelets (small fruits) in each berry and find the mean number. Use the same ten berries (or others) from each bush to extract the juice and test a standard amount by heating with Benedicts solution for ten minutes in a water bath. Consult *Nuffield Biology* Book 3. Prepare a colour standard, so that you can compare the percentage of reducing sugar in unripe and ripe fruits from the same bush and from a different variety or species of bush.

Look out for the following evidence of animal life:

1   The white winding tunnels of the larva of the bramble leaf miner the moth *Nepticula aurella*. These show up best on the old leaves. Make a collection of them, pressing them under clear adhesive tape. Study the mines carefully and find out all you can about the life history of this insect from books and from your own observations.

Serpentine mine made by caterpillar of micro-moth *Nepticula aurella* in bramble leaflet.

2   Nests of thrush, blackcap, robin and woodcock. Do not disturb nesting birds or touch or take the eggs. Old nests from last year may be taken and studied. See p. 86.

3   Sap sucking insects such as aphids and the foam-secreting nymphs of frog-hoppers feeding on the young shoots, together with various species of shield bugs.

4   Observe, describe and identify any pollinating insects. The wasp-like hover fly *Syrphus ribesii*, is one of these and its larvae resembling small transparent slugs live on the bramble leaves where they prey on aphids, eating about fifty a day.

5   The grubs of the raspberry beetle *Byturus tomentosus* and the raspberry moth *Lampronia rubiella* are the grubs found in the fruits. They can often be found by washing a jar of freshly picked fruit under the tap, and collecting the larvae which will be found in the water.

Hive bee on bramble flower.

6   Blackbirds feeding on the berries and wiping their beaks on the leaves to get rid of the seeds. Many other birds feeding on the berries and the insect life.

7   The common wasp preys on flies early in the summer and starts to feed on sugars in late August. It pierces the skins of the individual fruitlets to reach the flesh inside. Once the wasp has pierced the skin other insects swarm to the fruit including green bottles and flesh flies and butterflies such as commas, speckled wood and red admirals.

8   Spiders feed on the flies and their webs are very clear in the autumn mists and repay careful study.

9   Other animals that feed on the berries include the badger, the yellow coloured dusky slug and a number of snails, including the banded snail.

10   Around decaying tree stumps, seeds voided by birds may be seen together with young bramble shoots growing amongst moss and decaying humus.

Check sheet for bramble species

HABIT OF GROWTH

Branches arching, or non-arching

Rooting at tips, or non-rooting

ANGULARITY OF STEM

Rounded

Angled

COLOUR OF STEM

PRICKLES ON FIRST YEAR STEM (ARMATURE)

Position (on angles of stem, for example, irregular)

Size

Bramble stem, compound leaves and prickles.

Bramble from different species.

    Shape, (include samples stuck under clear adhesive tape)
    Direction in which they point
    Abundance
    Pricklets, acicles and stalked glands present or absent
    (See *British Flora* for definition of these and other terms.)

LEAVES
    Number of leaflets in compound leaf
    Nature and colour of hairy covering (indumentum)

INFLORESCENCE
    Shape
    Branch pattern
    Armature (prickles)
    Indumentum (hairs)

FLOWER
    Shape, Size, Colour

STAMEN
    Length relative to styles
    Colour of filaments
    Anthers (smooth or hairy)

STYLES
    Colour at base

77   (A)OE   The stinging nettle Urtica dioica *and the associated animals*

This plant is widespread, common and readily identified.

Distribution maps may be made to cover areas in the neighbourhood where nettles are abundant.

Starting in early spring when the first new shoots of this perennial plant appear, study the rate of growth by choosing ten plants at random (see Appendix note) from the same patch each week and measuring the height of each to the nearest centimetre. From the mean results each week construct a growth curve. Relate this to the rainfall and maximum and minimum temperature each week. Also compare rate of growth in different patches in different locations where the physical factors vary, for example patches in light and shade, in damp and dry situations. Relate this to light and humidity readings.

Dig up plants at each season of the year and study the growth of the rhizome system. Record date of flowering and examine flowers. Investigate the nature of the stinging hairs and find out all you can about them and their effect.

Every time you visit the patch to record growth rate, look for insect life, especially aphids in springs and the eggs, caterpillars, chrysalids and adults of the small tortoiseshell and red admiral butterflies in the spring and summer. Study these insects and their life histories in the field and in the laboratory to find out all you can about them and their relationship to the nettle plant.

Some plants may be found which bear galls on the flowers and leaves in summer. These are made by the gall gnat *Dasyneura urtica* and are worth investigation. (See Project 83.)

78   (A)OE   The wild arum Arum maculatum *and the associated animals*

This well known plant is also called 'Lords and ladies', 'Cuckoo Pint' and 'Jack in the pulpit'. It is widely distributed, common and variable with features of interest connected with its pollination mechanism.

Study the distribution of this plant in your neighbourhood. Map the position of the main patches on a large scale sketch map. If possible, start the study in early spring when the leaves appear and continue until the berries are dispersed in late summer and autumn.

Wild arum *Arum maculatum* in early May showing leaves, spathes and spadix.

Wild arum (spotted variety).

If you can locate say fifty or one hundred plants, find what percentage of these have spotted or unspotted leaves. Record the date at which you first notice   a) leaves,   b) spathes (the leaf like structures which enclose the flowers),   c) Flowers being pollinated by tiny moth-like flies,   d) fruits ripening,   e) all fruits dispersed.

Take light readings with an environmental comparator or light meter near each patch on a sunny day. What is the maximum light reading? Take a standard reading in the full sunshine for comparison.

In April or May when the flowers are pollinated, open several spathes, examine and draw the flowers and find out what you can about the mechanism of cross-pollination. Sources of information are Maud Jepson *Biological Drawings* and the *AA Book of the Countryside*. Capture a few of the flies and kill them in 100% ethanol. Draw under the stereoscopic microscope or make a permanent slide, using Hoyer's Medium (see Appendix).

Collect ten or more plants from different patches when the berries are ripe in July or August. How much do the numbers of berries vary between ten, twenty or more plants? Find out what you can about the dispersal of the seeds. Does the number of seeds vary from berry to berry? How much? N.B. THE BERRIES ARE POISONOUS.

This plant is a perennial with a tuberous rootstock. Dig up a whole plant in early spring and examine and draw the tuberous root. Carry out standard food tests on the root for starch, reducing sugar and protein. Devise a method to compare the amount of starch present in early spring and at the end of the summer.

Reproduction is both vegetative, by the creeping rootstock and sexual, by flowers, fruits and seeds. By means of observation try to find out which is the most successful method in this plant. Try to germinate the seeds,   a) as soon as you think they are ripe,   b) after keeping them in a cold place throughout the winter. Germinate them in small pots or seed trays with damp soil taken from where the plants grow.

79   OE   *A study of the yellow hill ant* Lasius (Acanthomyops) flavus

This is the ant which makes the mounds so commonly seen in meadows and pastures. Locate an area where they are common and show the position of each mound on an accurate scale map. Show the height of each mound in cm and work out the mean and the standard deviation. Mark the position of the steeper slope and record whether the mound is occupied or not. The summit nearly always faces south-east to catch the morning sun, the longer slope being to the north-west. You do not need to disturb the nests to see if they are occupied. If the surface of the hillock feels soft and spongy and gives way when you press it then it is occupied; if it feels hard and unyielding then it has been abandoned. *N.B. These ants often take over an old mole hill, so look out for this.*

Choose one or more occupied mounds to study in detail over a year. Measure each precisely and record the measurements. Repeat at intervals throughout the year. Each colony is said to bring up 26 oz of soil a year. Devise some way of testing this, converting to metric units. Test the pH of the soil in various places inside the nest and in soil taken from 1 m, 5 m and 10 m from the nearest mound. What do your findings show? Place a number of thermometers at different depths in the mound, taking care not to break them. Also stick one in the ground nearby. Repeat at different seasons of the year and account for your findings.

In early spring, slice through an occupied mound with a strong spade, disturbing each half of the mound as little as possible. Insert a piece of red perspex between the two halves of the mound. Carefully shovel away one half, placing it, or part of it, in a strong polythene bag. This portion can be taken back to base for further examination and study or placed to one side to see what happens to it during an interval between visits. Push the perspex down into the ground and you will find that as the ants are insensitive to red light, you can study the structure of the nest and the activity of the ants. If the perspex is not likely to be disturbed it should be left in position for observation on subsequent visits. If, as is most likely, it cannot be left in place, repeat the exercise in summer and winter using different mounds. Red perspex can be used during your observations inside as well. If you intend to keep the ants inside for any length of time, read about how to keep them in a formicarium in Wragge-Morley *The Ant World* or one of the other reference books in the Further Reading List.

Fields (Projects 80–85)

80   A(O)   Use of exclosures in fields to compare growth of vegetation inside and outside them

This project is designed to be carried out on a farm in a field with cattle or sheep and obviously this must be in full co-operation with the farmer! The purpose is to find out the effect of grazing animals on the vegetation.

Choose a part of the field a few metres square say 2.5 × 2.5 m and erect a cattle or sheep proof fence around it. The area should be typical of the field as a whole, and not by the hedge or in a sheltered corner. It is best to set up the exclosure in early spring before the vegetation has grown too much. Set up rain gauge and maximum and minimum thermometer nearby.

By random selection (see Appendix) select a half metre square within the exclosure and one outside it, but near it. Fill in details of the vegetation using the method given in the Appendix for use of the quadrat. If plants cannot be named, and this is often difficult before flowering, take samples of each plant, from outside the study area if possible, press them and stick them under clear tape giving them code names for future reference (see p. 12).

From each half metre square, select a decimetre square at random, cut the vegetation as low as you can and put each sample in a separate labelled polythene bag. Record the rainfall and maximum and minimum temperatures each week and repeat the cutting operation, selecting the squares at random each week both inside and outside the exclosure.

Back at base, weigh the cuttings, record the fresh weight, dry completely in a low oven or incubator. Reweigh and repeat heating and weighing until two weights are the same. Record this dry weight.

Make bar graphs to record growth of vegetation throughout the season and try to relate it to the weather conditions and to the feeding activity of the herbivores. Remember that other herbivores, for example rabbits and insects may be eating the vegetation both inside and outside the exclosure. Try to find out something about the feeding activities of these other animals.

In late summer try to make a more complete identification of the plants both inside and outside the exclosure and see if there appears to be any selection of plants by the cattle or sheep. Which plants survive best under the different conditions?

81   A.   Grasshopper population study

A large collecting net, about 40–50 cm in diameter, on a stout frame and made of strong nylon is required. No handle is needed. It should be possible to get the head and upper part of the body and arms inside this net to sort the grasshoppers and mark them. Also required is a large open necked jar and lid into which the grasshoppers can be put temporarily, a killing bottle (see Appendix) and a paint brush, and a small tin of waterproof enamel paint.

Preliminary searches must be made with the aid of the net, to find an area of grassland in which there is an established population of grasshoppers. Sweep the net strongly amongst both short and long grass, keeping the mouth of the net shut by turning it over at the end of the sweeping sessions to prevent the escape of any insects caught. Examine the catch from time to time to see whether any grasshoppers are present. Failing grasshoppers, this project can be adapted for other species of insects which are abundant. Take a small sample of the catch back to base in the killing bottle and identify the insects caught.

If you have caught ten or more of any particular species from a small field, use the species for the population study.

Restrict the population study to a well defined area where it is likely that the insects chosen will be fairly evenly distributed, say grasshoppers in a small field where the conditions are uniform throughout.

Carry out a certain number of standard sweeps or sweep for a chosen length of time, say half an hour, covering the whole field. Mark each grasshopper or other chosen species of insect by a small dab of paint on the dorsal side of the thorax. It may be easy to do this in the net or possibly after transfer of the grasshoppers into a wide necked jar. Count the insects marked and release them as soon as the paint is dry.

Repeat the catching procedure under similar weather conditions and at the same time of day, a few days later. If you catch none of the marked insects it will be necessary to mark another sample and repeat the procedure until some marked insects are caught. The total population is calculated by using the formula called the

Lincoln Index, as follows:
  Where S_1 is the number marked in the first sampling
  S_2 is the total number in the second sampling
  S_3 is the number of marked individuals in the second sample
  N is the total population

$$N = \frac{S_1 \times S_2}{S_3}$$

  This formula holds good for all freely mixing and non-territorial animals.

82   A(O)   Colonisation of cow pats

For the non-squeamish, this project is highly recommended for the insight it gives into plant and animal succession and the concept of the microhabitat.

  Find an easily accessible field where there are cows. Examine the cow pats to determine whether they are new or old. It may be possible to classify them according to the following stages   a) steaming hot,   b) cooling, no crust,   c) cold, crust beginning to form,   d) thick crust, but liquid beneath,   e) very thick crust, solid throughout,   f) only just recognisable owing to almost complete decomposition and incorporation into the soil. Photographs of the various stages would be useful.

Cowpat at stage b) Several dung flies just visible.

Cowpat at stage e).

Examine several cow pats at each stage and record any plant or animal life visible on the surface. The plants will be fungi in the last three stages and possibly mosses and flowering plants in the last two stages. The animals will be mostly beetles and flies and evidence may be seen of birds pecking at the pats to extract the insect larvae. Record all that you see, making sketches and notes. Use an old chisel and trowel to move away the surface carefully and expose the larvae and beetles below the surface. Take samples back with some of the substrate in labelled, adhesive taped petri dishes, or in jam jars if you wish to hatch out the larvae.

The most notable beetles are the Scarabaeid beetles and their larvae, especially the Dor beetle *Geotrupes*. The female beetle burrows about 40 cms below a batch of dung, filling the blind end with a plug of dung which is food for the larva. It is possible to bring the adult beetles up to the surface by damping it with water.

The yellow dung-fly *Scatophaga sterocaria* is another abundant scavenger which frequents patches of fresh cattle dung in large numbers in summer. The eggs are laid in the dung, where the larvae feed and come to maturity.

Amongst the other fly larvae and beetles present, some are dung eaters but others are predators, keeping the first group in check as far as numbers are concerned.

Although some work on the more obvious, larger fungi can be done in the field, it is best to bring samples of cow pat to the laboratory to incubate in order to see the full succession. Dung will always produce characteristic fungi whatever time of year it is collected.

Bring back samples of fresh cow pat in closed plastic containers and incubate them on damp blotting paper, placing the containers in a warm place. Earthworms and insect larvae should be removed from the samples if possible or their activities should be limited by spraying with an insecticide of the type used against flies.

Keep the dung under constant observation during incubation. The first fungi to appear are the moulds which are numerous and require a microscope for identification. Next, follow a series of Ascomycetes which are best seen with a powerful hand lens or stereoscopic microscope. The fruit bodies of the Basidiomycetes are readily seen with the naked eye. These are of the toadstool type.

Watling, in his book on the *Identification of the larger Fungi* lists twenty-six species or genera of dung colonising fungi and has also, with Richardson, compiled a useful key.

The methods given above can be adapted for other types of dung, including rabbit pellets, which can be incubated in covered petri dishes.

Fungi, *Coprinus* species on pony droppings.

83 AO(E)   Plant galls and their animal inhabitants

The abnormal growths of plant tissues called galls are caused by various kinds of living organisms. Most are caused by insects, some by mites and a few by fungi and eelworms.

Most people know the hard brown marble-gall and the rosy oak apple, but there are hundreds of other galls, mostly on oak but also on a great variety of other plants. Galls are vegetable growths entirely produced as a result of the stimulus of the irritant insect, fungus or disease organism.

Anyone interested in this project is advised to consult A. Darlington *Pocket Encyclopaedia of Plant Galls* and also to read a general account of their biology such as is found in J. Imms *Insect Natural History* and R. Lulham *Introduction to Zoology through Nature Study*. There is also much useful information in Molly Hyde *Hedgerow Plants*.

As the greatest number of galls occur on oak, it may seem strange not to put this project in the woodland section, but it is likely that the greatest number of different types of gall will be collected in hedgerows or in the countryside in general. As well as oaks, willows, birch (witches' broom), poplar, lime and rose should be examined for the typical galls which are described in the books mentioned.

For a short term project, collect as many galls as you can during both country and woodland walks as well as in the garden on plants such as blackcurrant (big bud). Identify and draw the galls on the buds, leaves or stems. Examine for exit holes of the gall-making animal and cut open to find the larvae and the parasitic insects which may be found with them. Identify the insects with the help of books. The best time for this work is probably late summer or autumn, but the oak apple is found in early summer.

If more detailed work or a longer project is desired attempts can be made to breed out the insects from the galls in jars containing sand and covered by muslin. Patience is needed and a good hand lens or microscope as the insects and mites responsible are very small. Many of the insects which emerge from the galls will be parasites of the gall-causers or iniquilines which feed on the gall substance and have their own parasites and hyperparasites. Draw the insects and preserve them or make permanent slides as described in the Appendix.

84   (A)OE   *Variation in the number of floral parts in some common plants*

Hundreds of different species of flowering plants may be found in the three habitats under consideration. Of these many are common and easily identified. Each species shows considerable variation some of which is environmental and some of which is genetic. Of the genetic variations those concerning number of floral parts are probably the easiest to study and the least influenced by environment. If the study is sufficiently detailed it can be subject to statistical analysis as well as to graphical illustration. The plants chosen here are common and easily identified, but many others would serve just as well in their place.

84.1   *Petal number in the lesser celandine* Ranunculus ficaria

This plant is found in the hedgerows and waste places in early Spring. The usual number of petals varies between eight and twelve. Collect one hundred flowers and count the number of petals in each. Express as a bar graph. This might be done in several different localities or in the same locality at weekly intervals during the flowering period. Do all the graphs show the normal bell shaped curve of a continuously varying factor?

84.2   *Stamen and carpel number in the lesser celandine*

This can be done in a similar way although the counting is more tedious. It is sometimes said that botanists cannot count beyond ten! Floral formulae, giving the number of parts in a flower, count any number above 10 as numerous. By counting the stamens and carpels of this species it may be possible to recognise two distinct morphological forms of the plant as the variety *fertilis* has stamens 9–60 and carpels 11–72 and is the commoner type, especially in sunny places. The variety *ficaria* has stamens 14–40; carpels 5–44, and is a more local plant, chiefly in the shade. Devise a way of distinguishing between these two varieties by a series of bar graphs. Both in 84.1 and 84.2 your work will be more interesting and attractive if you stick in samples of the counted floral parts under clear adhesive tape.

84.3   *Stamen and carpel number in three species of buttercup*

Project 84.2 described above can be carried out with any of the three common species of buttercup *Ranunculus repens* with creeping stems, rooting at the nodes, *Ranunculus bulbosus* with reflexed sepals and *Ranunculus acris* with neither of these features. For other differences see *Flora of the British Isles*.

Carry out the project as in 84.2 above using one, two or three of these species and looking out for any differences between the bar graphs obtained.

84.4   Floret number in the daisy Bellis perennis *or the dandelion* Taraxacum officinale

All members of the Compositae family, to which these plants belong, have composite flowers each consisting of an inflorescence or collection of tiny flowers, so that each apparent 'petal' of the daisy is a tiny flower or floret. Carry out the project in a similar way as described above. It might be interesting to see how many other members of this family you can identify and take ten inflorescences from each to count the florets.

84.5   Number of flowers in the inflorescence of members of the Umbelliferae or Labiatae family

If counting the florets of the daisy or dandelion is too tedious, choose plants like the cow parsley (Umbelliferae) where the flowers are numerous but slightly larger or the deadnettles (Labiatae) where the flowers are fewer and larger. The same principle applies.

85   OE   Fruits, seeds and dispersal

All three of these habitats yield a rich harvest in late summer and autumn and the possible number of projects is immense. As any elementary or O level examination syllabus usually contains this subject, a class project is given here, but any part of it can be taken up as an individual or small group project.

If the school is within easy reach of suitable habitats the project can take place in usual lesson time, if not then a half day expedition is recommended.

Divide the class into six groups studying between them three groups of plants, A, B and C. Thus a certain amount of competition is introduced as two groups will be studying A, two groups B, and two groups C.

Group A plants include ash, sycamore, maple, hornbeam, pine, elm, thistle, willowherb, dandelion, groundsel, wild clematis and any other winged or plumed fruit or seed that may be found locally.

Group B plants include bramble, elderberry, wild rose, hawthorn, black bryony, white bryony, sloe, crab apple, snowberry, holly, wild geum, burdock, goosegrass, agrimony, enchanter's nightshade, oak, horse chestnut, beech, hazel and any other animal dispersed fruit that may be found locally.

Group C plants include broom, gorse, vetch, clover, bluebell, scarlet pimpernel, plantain, figwort, campion, poppy, horse chestnut, shepherd's purse and any other with explosive type fruits that may be found locally.

Before going out, the members of each group should examine pictures and drawings of the plants concerned, or slides may be shown to the class.

Choose a leader for each group to co-ordinate the collecting. Polythene bags may be used, but an open chip basket is better for the succulent fruits. Collect a few examples of each type of fruit. For the plant underlined in each group, collect one sample per person or more if possible (about ten). For the rose also count the number of hips on one plant.

Fill in a table similar to the one set out below. Books may be needed and Maud Jepson *Biological Drawings* is recommended.

Name of plant	Seed, fruit or collection of fruits?	No. of seeds in fruit	Agents of dispersal	Parts used in dispersal
pine	winged seed	—	wind	expanded
bramble or blackberry	collection of fruits	one in each small fruit	birds and mammals	black skin or epicarp and juicy mesocarp
rose	false fruit	one per true fruit	birds and mammals	bright red receptacle
vetch	fruit	several	—	drying and twisting of pericarp valves

etc.

Each person in the group should choose a different plant and make a much enlarged drawing to show all the features used in dispersal. Stick all the drawings from one group on to a large poster and display with the specimens. Study these and those from other groups. Include a list of the advantages and disadvantages of the particular method of dispersal illustrated by the plants in your group.

For the dandelion, rose and plantain and other plants, if possible, calculate the number of seeds produced by each plant. Study each of these plants in more detail in the field and laboratory to assess the efficiency of the method of dispersal. Why do some seeds fail to germinate after dispersal? See *Nuffield Biology* Book 3 for some ideas for experiments and devise others of your own.

Hedgerow projects (Projects 86–88)

86   OE   Composition of hedgerow

86.1   Age of hedgerow related to number of plant species it contains

86.2   Birds and insects of the hedgerow

86.3   Dangerous litter
The RSPCA have published a folder containing information cards, project cards, booklet and cards for return of information on these three subjects. See also Molly Hyde *Hedgerow Plants* for all the hedgerow projects.

With our fast disappearing hedgerows it is important that we understand their history, maintenance and ecology. At present they form refuges for wildlife driven out of other habitats by changes in farming patterns, destruction of woodland and building activities.

These projects are highly recommended for younger children and those with little background knowledge.

87   (A)OE   Adaptations of plants to their position in the hedge and climbing plants
Choose an interesting stretch of hedgerow which contains many different species of woody and non-woody plants, including

climbers. It is best if both sides of the hedge are accessible.

Observe and record the physical features such as aspect (N, S, E or W) maximum and average height, maximum and average width, presence or absence of ditch, whether bordering field, lane or road, temperature and humidity at different heights and on both sides at different times of day and at different seasons. Temperature can be measured with a maximum and minimum thermometer if this can be safely hidden. Humidity can be measured by timing the rate at which blue cobalt choloride paper turns pink, or more accurately by a wet and dry thermometer (see Appendix).

Identify the shrubs which compose the hedge, the climbers and the non-woody plants or herbs. If there is a ditch at the base of the hedge, aquatic or semi-aquatic plants may be present. Relate position of plants to physical factors if possible. Plants with weak stems tend to be climbers. Some climb over the grass and herbs at the base of the hedge, like purple vetch, or over the shrubs, for example black bryony. See how many of these climbing plants you can identify and make careful sketches to show which part of the plant is adapted for climbing. Some plants, such as vetch, have tendrils formed of leaflets. Others, such as white bryony, have petioles (leaf stalks). A third group, like black bryony and the larger

Weak stemmed goosegrass using clinging hairs to climb in hedge.

bindweed or convolvulus, have twining stems. Still others, like the ivy, have clinging adventitious roots, growing from the stem.

Keep records of the hedge throughout the year, if possible recording dates of leaf opening and leaf-fall of deciduous shrubs, flowering and fruiting. Record insects seen pollinating flowers, adaptations to cross-pollination by insects or to wind pollination. Note adaptations to wind, animal and explosive methods of dispersal of fruits and seeds. Investigate methods of vegetative reproduction in some of the hedgerow plants.

88 OE Census of common birds
This census could be carried out in woodland or in meadow land with hedges by students who can recognise at least ten species of common bird accurately by sight and sound. Choose an area which you can visit frequently which is covered by one 25 inch Ordnance Survey map or plan. It is usually more difficult to spot birds in woodland.

Make ten visits to the area from April to the end of June. Plot the position on the map of any birds which you see or hear. Transfer the information on to copies for individual species. At the end of the exercise clusters appear which indicate the territory of individual birds. The total number of clusters for all species gives some idea of the bird population of the area.

Combine the census with further study of the habits of the chosen species to make a more ambitious project.

Further Reading

1   AA Drive Publications *Book of the British Countryside* 1973
2   Skytte Christiansen, K. *The Pocket Encyclopaedia of Wild Flowers* Blandford 1965
3   Clapham, A. R., Tutin, T. G. and Warburg, E. F. *The Flora of the British Isles* Cambridge 1962
4   Connold, E. T. *British Vegetable Galls* 1901
5   Connold, E. T. *British Oak Galls* 1908
6   Darlington, A. *The Pocket Encyclopaedia of Plant Galls* Blandford 1968
7   Darlington, A. *Field and Meadow Life* Blandford 1973

8   Fitter, R. and A. and Blamey, M. *The Wild Flowers of Britain and Northern Europe* Collins 1974
9   Frost, S. W. *Insect Life and Natural History* Dover Books 1959
10   Gilmour, J. and Waters, M. *Wild Flowers* Fontana New Naturalist 1972 (Useful Key to common Umbelliferae and Compositae)
11   Hyde, Molly *Hedgerow Plants* Shire Publications 1976
12   Jepson, M. *Biological Drawings* Murray 1942
13   Lulham, R. *An Introduction to Zoology through Nature Study* 1923
14   Martin, Keble *The Concise British Flora* Michael Joseph 1965
15   Phillips, Roger *Wild Flowers of Britain* Pan Books 1977
16   Prime *Lords and Ladies* New Naturalist Collins
17   Pollard, E. Hooper, M. Moore, N. *Hedges* Collins 1974
18   Royal Entomological Handbook *Ants* R. E. Society 1975
19   RSPCA Education Dept., *Hedges*
20   Swanton, E. W. *British Plant Galls* 1912
See also titles on Further Reading Lists: Chapter 1 No. 6; Chapter 2 No. 10; Chapter 4 Nos. 4, 5, 6, 7, 17, 19, 26, 28, 30, 31, 37, 41, 44, and 45; Chapter 6 Nos. 4, 5 and 9.

6 Town and Garden

Many children or young students who wish to carry out fieldwork projects live in towns and may find it impossible to attempt any of the long term projects in the earlier chapters of this book. Some of these can be adapted to a garden or park habitat, for example:

Chapter 2   Freshwater Projects 34 and 36
Chapter 4   Woodland Projects 52, 65, 66 and 68
Chapter 5   Field, Hedgerow and Wasteland Projects 77, 84 and 85

Others are ideal for a holiday project, such as many of those in Chapter 1.

This section includes a number of projects which can be done in park, garden or even the house. They qualify as fieldwork projects as they involve working outside the classroom or laboratory and provide opportunity for research by individuals and small or large groups.

Town and Garden Projects

AOE	89	Plant and animals associated with walls
(A)OE	90	Air pollution projects
	90.1	Distribution of Pleurococcus and Lichens
	90.2	Behaviour of privet
	90.3	Pollution in home, school and garden
AOE	91	The London pigeon—variation and habits
OE	92	The London plane tree
AOE	93	Leaf mining insects in common garden plants
OE	94	Insects and other invertebrates trapped i) accidentally or ii) intentionally
OE	95	Weeds in the lawn
OE	96	Earthworms in the lawn
AOE	97	Pollination—observations and experiments
OE	98	Garden birds
AO(E)	99	A study of web-making spiders in the house and garden
AOE	100	Habits of garden molluscs
A(OE)	101	The ecology of the bird bath with special reference to the rotifer *Philodina roseola*
OE	102	The ivy *Hedera helix*
(A)OE	103	Annual and perennial weeds in the garden

89 AOE Plants and animals associated with walls

Most walls, even in industrial cities, support some kind of plant life such as the ubiquitous alga *Pleurococcus* and a few mosses and lichens. *Nuffield Biology*, Book 3, outlines a project which studies the distribution of *Pleurococcus* and this can be adapted for lichens and mosses.

First explore your neighbourhood and notice whether the walls are brick or stone, old or new, with few or many plants. Include only the plants which grow on the wall surface or have roots in the wall itself. If there is a castle in the town or an old church these may be especially rewarding to study.

Pellitory of the wall *Parietaria diffusa* in church wall.

Decide whether you will study walls in general, brick and stone walls compared, or one particular wall in detail. This will depend partly on the plants which you find, but also on the amount of time you can spare.

Plants which grow on walls include those which are also common elsewhere, for example *Pleurococcus* and *Urtica dioica* the stinging nettle. Others are found only on walls, such as the fern called wall rue *Asplenium Ruta-muraria* and the flowering plant, pellitory of the wall *Parietaria diffusa* (formerly *officinalis*). Identify the plants

White crustaceous lichen on very old wall at Dunvegan castle. (Skye)

Lichens and naturalised daisy *Erigeron mucronatus* on old brick wall.

with the help of the *British Flora* or other wild flower books given in the references. Make sketches or photograph the walls and plants to show the exact position they occupy.

Wall pennywort *Umbilicus rupestris* on dry stone wall.

Ivy-leaved toadflax *Cymbalaria muralis* on dry stone wall.

Study the adaptations of the plants to this usually very dry and unpromising situation. These include devices for storing water in succulents like wall pennywort *Umbilicus rupestris* (formerly

Cotyledon umbilicus), or reduction in leaf surface as in procumbent pearlwort *Sagina procumbens*. In the ivy-leafed toadflax *Cymbalaria muralis* (formerly *Linaria cymbalaria*) the stems carrying the ripe fruit become negatively phototropic, thrusting the seeds into crevices of the wall.

Lichens and pennywort (young plants) in stone wall with mortar.

Observe and investigate carefully the structure of the wall. How is the mortar converted into soil over a period of time? What part do mosses, lichens and weather play in this? Analyse the soil from older walls for humus, water and lime content. (*N.B. mortar is about 70% lime.*) Compare the content of nearby garden soil.

Compare the vegetation on different sides of the same wall. Take light readings at different times of day. Compare exposure to wind and rain. Relate these factors to the plant life.

It is not so easy to study the animal life of walls as it is likely to be sparse compared with the plant life. It is worth scraping off samples of *Pleurococcus* and various lichens from walls and seeing if they support any microscopic life. A minute rotifer, or wheel animal, is sometimes found in association with *Pleurococcus*. Collect samples of soil from the crevices in walls and examine with a hand lens for animals such as snails, spiders, worms, wood lice, beetles, centipedes and millipedes. Samples may also be analysed by the use of the funnel method given on p. 73. See also p. 136 for spiders associated with walls.

90 *(A)OE Air Pollution projects*
The Pollution Handbook by Richard Mabey describes three such projects which can be used as described or adapted and extended, as indicated below.

90.1 *Distribution of* Pleurococcus *and lichens in relation to air pollution*
This is described in detail and need not be elaborated here.

90.2 *Behaviour of privet in relation to air pollution*
This is only described briefly with a table giving some results. In the cleanest air privet is evergreen and flowers, while in the most polluted air it drops its leaves in November and does not flower. The white flowers should be looked for in June and July and the black fruits from June to August. Survey the chosen area during these months and record the information on maps and in tables. Repeat in November, December and January to check on leaf fall. This may be partial or complete. Scrape off the grime deposit from the leaves in autumn and calculate its weight in grams per square metre of leaf surface (or weigh leaves before and after washing).

Carry out experiments on the rate of transpiration of a privet shoot at different times of year and from different places at the same time of year. This can be done using a potometer, in which case each shoot must have the same number of leaves or the results converted to rate per leaf. Alternatively enclose shoots in polythene bags over a period of time and measure the weight of water collected in a week. Allowance must be made for differences in weather condition and number of leaves. Air pollution may reduce the rate of transpiration to as little as one tenth of that in unpolluted areas.

90.3 *Pollution in home, school and garden*
This can be measured by the use of white card on which a number of rubber bungs or children's bricks are placed, one being removed

each day. Use of this method is considerably more difficult outside than inside but imagination and ingenuity should provide a way. Spray the results with a pastel fixative to make them permanent and repeat the method in as many places as possible.

91   AOE   *The London pigeon—variation and habits*

The London pigeon lives mainly in fully built-up areas of towns, but rarely in suburbs with gardens. It occurs in several plumage types, with much intergrading between them. Blue rock has prominent black bars on the wing, no black wing tips and a prominent white rump. Blue chequer has whole upperparts except tail and white rump mottled grey-blue and black. Red rock is cinnamon-red with white wing tips and no black bars. Red chequer is similar but with wings and mantle mottled white. Black forms occur and others are white or much splashed with white.

Students living in London may choose to study the pigeons in one locality such as Trafalgar Square or St James's Park or make a comparative study of pigeons in many squares or parks. Those living in other large towns might compare various areas or study one in area detail. This project could occupy one day, one week, one month, one year or longer.

Make out a table with the six types given above as headings and columns for other types seen. Take a supply of crumbs and a partner to help recording. Sit on a park bench and throw down the crumbs. Record the numbers of each type of pigeon seen within a chosen time, say fifteen minutes. Repeat in other parts of the park/square and in other parks. Make notes on tameness and habits (feeding and display).

Choose one very distinctive type and try to keep it under continuous observation for as long as possible. A small portable tape recorder is useful for this. Make sketches or take photographs of the six varieties and intermediates.

Find out what you can by direct observation and enquiry about the pigeon as a public nuisance and the steps taken by Local Authorities to deal with it.

Be careful to distinguish from other closely related forms such as wood pigeon and stock dove. Pigeons occurring in mixed flocks of many different colour varieties are likely to be London pigeons.

92 OE The London plane tree—variation and adaptation to London life

This tree is as typical of London parks and squares as the London pigeon. While the bark of most London trees is black with grime, that of the plane peels off in patches, leaving a smooth yellow patch behind, which is free from contamination. Choose a tree where patches of different coloured bark are easily accessible. Mark a patch, several decimetres square, in some permanent but unobtrusive and harmless way. Use a transparent decimetre square quadrat frame marked in centimetre squares. Record the pattern and colour of the bark on graph paper. Use pieces of thick plain paper cut to the same size as the quadrat. Stick with adhesive tape on to the tree and rub firmly across the whole surface to obtain a grime pattern. Compare with the colour patterns. Does the oldest bark give the darkest colour? Collect all the bark which has fallen off the tree recently. Dry it and weigh it and stick samples under clear adhesive tape.

Repeat all these procedures at different seasons of the year in order to answer these questions. How much does the bark pattern change over the year? Does the grime pattern match it? At what season of the year does most bark flake off the tree?

In autumn, collect about one hundred fallen leaves from several different trees. Keep each batch in separate polythene bags. Measure the length and width of each leaf, standardising the method you use. Show the results by means of bar graphs. Do the different trees show similar results? Choose the largest leaf and draw round its outline. Select ten other leaves at random (see Appendix). Draw their outlines inside the largest one, using a different colour for each. Are any of them exactly alike? Examine the rest of the leaves from the same tree. Can you divide them into several groups of distinct shape? If so, give a name to each pattern and find out how many there are of each kind. Repeat with the leaves from the other trees. Does each tree show a similar size range and pattern range? Are the size and pattern variations examples of continuous or discontinuous variation?

In autumn and winter collect the round fruit balls as they fall from the tree. Collect a hundred from each of several trees in separate bags. Select twenty fruits at random from each bag. Count the number of separate fruits in each ball and represent the results by a bar graph for each tree. If you can count all the fruits the results will be more accurate. Find out what you can about the method by which the seeds are dispersed.

In summer ask the park keeper to cut you a spray of leaves on a piece of woody twig which will fit into the rubber or plastic tube of a school potometer. Obtain this after a spell of dry weather. On return to base, cut an inch of the twig under water and set up the potometer in the standard way. Find the rate of water uptake (proportional to the rate of transpiration) over a suitable length of time. Next, sponge the leaves very carefully to remove the grime which may block the stomata. Dry the leaves and repeat the experiment. Do the results support the information that rain showers wash the leaves clean so that transpiration is not greatly hindered by pollution?

93   AOE   Leaf mining insects in common garden plants

Mention was made in the previous chapter of the bramble leaf miner. Many common garden and wild plants also contain leaf miners. These may be the caterpillars of tiny moths as in the bramble and rose, which form serpentine mines. Those in the hawthorn, lilac, ash and hazel form blotch mines. Fly or dipteran

Leaf miners in sow thistle.

larvae forming leaf mines may be distinguished from the moth larvae by their more cylindrical shape and their lack of legs and head. The chrysanthemum leaf miner forms serpentine mines in this plant and also in many other members of the Compositae family such as marguerites, Michaelmas daises and weeds like the sow thistle. The holly-leaf miner forms blotch mines, as do celery fly larvae in celery and many of its wild relatives. Other common garden plants containing leaf miners include nasturtium, dahlia and honeysuckle. If possible consult *The Natural History of the Tineina* (1855–73, 13 vols) in the British Museum of Natural History for beautiful hand coloured plates of the mines made by moth larvae. A more general book is Needham, Frost and Tothill, *Leaf-mining insects*, while other useful books are J. Imms *Insect Natural History*; A. Darlington *Woodland Life* and S. W. Frost *Insect Life and Natural History*.

Serpentine mines in honeysuckle *Lonicera periclymenum* leaves.

Collect as many examples as you can of leaves attacked by leaf mining insects, during summer and autumn. Examine the mines carefully. What part of the leaf do the larvae eat? How many are there in each leaf? Do they cross the mid rib or the smaller veins? Notice how serpentine mines grow wider as the larva grows and how patches of dots mark the excretory matter or frass produced from the undigested food. Stick an example of each mined leaf under clear tape. Remove the larvae responsible wherever possible and classify each as lepidopteran or dipteran. Draw and preserve for future reference (see Appendix). Try to breed out the adult insect by keeping a spray of leaves which are known to contain larvae in insect cages (described on p. 84) with damp soil or sand in the bottom. Some larvae pupate in the leaf and some in the soil. Change the water in the container often and look out every time for signs of the tiny adult moths or flies which may emerge.

94   OE   *Insects and other invertebrates, trapped accidentally and intentionally*

i) ACCIDENTALLY

The idea for this project came when a family with small children occupied my flat while I was away. The landing light had been left on all night for a week and window open. On my return I found the bowl fitting on the light bulb full of an assortment of night and day flying insects. When these were removed and mounted on card they proved very useful for illustrating the classification of insects and the variety of adaptations which they show. Dead insects are often found in certain shaped reading lamps and fluorescent fittings during or at the end of a long hot summer. Insects are also attracted to sun rooms and conservatories containing plants. They may be overcome by the heat during the day and litter the window-sills and floors. Instead of sweeping them up for the dustbin, collect them each day, stick them on card and relate the numbers and types found to the season and the weather conditions inside and outside. Draw and identify as many as possible. Find out what you can about their habits from observation in the garden or from books.

Other insects may be trapped accidentally in the garden itself, on the surface of ponds, water butts or even puddles. Collect these, dry them on blotting paper. Mount and examine as described above.

ii)   INTENTIONALLY

Two kinds of traps which are very simple to use in the garden are the pit-fall trap for ground-living insects and the water trap for flying insects. If used discriminately these provide useful information without any profound effect on the garden fauna.

Pit-fall traps consist of 1 lb jam jars baited with a very small portion of raw meat. Sink them in the ground so that the top of the rim is level with the ground surface. Place several in different parts of the garden and examine each morning for captures. These are likely to include beetles, spiders and slugs. Describe and/or draw each animal before liberating it. If you are a vegetable grower you will no doubt prefer to destroy harmful insects, snails and slugs or kill and preserve them for future reference.

Water traps can be plastic or enamel dishes with a large surface area to depth ratio. Large enamel developing dishes are ideal. Half fill these with water and place strategically in the garden. Try the effect of painting them different colours on the numbers and types of insects caught. Examine daily or weekly for catches. Remove the insects from the water carefully with a fine paint brush and dry with blotting paper. If large, mount them on card. If small, make permanent slides (see Appendix). Describe, draw and identify later.

With all these methods, the purpose is not just to make a collection, but to discover the range of insect and other invertebrate life in the garden and from this to study habits of the insects and to build up a picture of their part in the garden ecosystem.

Slugs and snails may be collected live under flowerpots and half orange or grapefruit skins. Alternately they may be collected by the use of commercial slug bait in the form of pellets. The latter method has the advantage of killing the slugs which will be appreciated by gardeners. Be sure to wash carefully after using the pellets or handling the dead animals. A useful book for identification is Michael Chinery *The Natural History of the Garden*. This book also gives many other ideas for projects, as does *The Family Naturalist* by the same author.

95   OE   *Weeds in the lawn*

This project can be carried out on the lawn at home or a lawn or playing field at school with the permission and co-operation of the gardener or groundsman. A moderately neglected lawn, or one where no chemical treatment is given regularly, is better than a prize-winning one for this study.

Use the quadrat frame described in the Appendix and select one or more half metre squares which contain plenty of weeds such as plantain, daisy, buttercup, dandelion. Record the plant(s) in the quadrat at monthly intervals, being sure to include some records before and after lawn cutting.

In the spring or early summer, carefully remove all the weeds from one or more of the patches, extracting as much of the root as possible and disturbing the grass and lawn surface as little as possible. Mark the position of the bare patches accurately and record the subsequent growth of grass and weeds at fortnightly intervals. Use the weeds you dig up to make careful observations, descriptions and drawings or photographs, noting the features which adapt the plant to success as a weed on the lawn.

In other patches try the effect of various chemicals on the different species of weeds. Try a pinch of kitchen salt placed on the centre of each plant and watered in later if it does not rain. Try the effect of various proprietary lawn weedkillers, following carefully the directions on the packet. Record your results in quadrat and bar graph form.

If possible try the effect of different types of lawn mower and mowing at different intervals on the weed growth. Firstly, use the close cutting cylinder type of mower and, secondly, a rotary mower which can be of the Flymo* type. Try also the effect of trampling on species distribution.

What conclusions do you reach about the best conditions for growth of the various species of weeds and the best treatment to get rid of each species?

96   OE   *Earthworms in the lawn*
This project can be combined with the previous one or carried out separately.

Select half metre squares as in the last project, choosing several which contain numerous worm casts. These will probably be most obvious in spring or autumn. Mark the position of each worm cast on the quadrat record and repeat at monthly intervals. Record in various

---

*N.B. the Flymo mower should not be used by children themselves.

weather conditions, noting these and the previous day's weather. Make use of a maximum and minimum thermometer and rain gauge if possible.

After one month of recording, carefully remove all worm casts from one or more of the quadrats. Place a very small quantity of this material on a glass slide with a drop of tap water. View with a hand lens or under the low power of the microscope. Draw a few of the particles to show their size and shape. Repeat with a sample of ordinary soil from the lawn. Record the return of worm casts to the cleared quadrat(s) at weekly or more frequent intervals.

Water one or more of the quadrats with dilute potassium permanganate solution (25 g to 4 litres of water) or 2% formalin solution. Count the number of worms that emerge from their burrows and choose a large one for identification and dissection. The long worm *Allolobophora longa* makes worm casts above the ground but does not plug its burrow. *Lumbricus terrestris* may also be obtained by this method. This species makes casts below the ground and plugs its burrow. For further help with identification and ideas for projects, see *Nuffield Biology* Book 1 (Teachers' Guide).

Kill the animal for dissection in boiling water, removing it as soon as it is still. See zoology textbooks or *Dissection Guides* (The Invertebrates) for dissection instructions. Identify and open up the different parts of the alimentary canal. Try to find out in which portion of the gut the large soil particles are ground up into smaller ones.

Relate your findings to soil fertility and to structure of the lawn surface.

97   AOE   *Pollination—observations and experiments*

These can be carried out in existing gardens, specially planted gardens or flower beds or even in window boxes and flower pots. A detailed account of such experiments is to be found in Lilian Clarke *Botany as an Experimental Science in Laboratory and Garden.*

The larger the range of plants the better. Some of the following plants should be present or planted specially: buttercup, broom, borage, dead nettle, foxglove, figwort, larkspur, gorse, monkshood, sweet pea, pinks, snapdragon, salvia. If there are bee-hives nearby this is ideal.

Find out if pollen is necessary for the formation of fruit by cutting out the stamens when the flowers are in bud and tying the buds in muslin bags. Count the number of fruits formed. Cut out the stamens when the flowers are in bud, apply pollen from other flowers of the same kind to the stigmas when they are ripe. Count the number of fruits formed.

Enclose the buds of as many different species of flowers as possible securely in muslin bags. Count the numbers of fruits produced to see if insect visitors are needed for pollination.

Be aware of possible sources of error in the above experiments and avoid them as far as possible or take them into account when writing up results.

By careful observation determine which species are visited by hive-bees, bumble bees, hover-flies, wasps, butterflies, moths and flies.

Take note of the number of flowers of the same species visited by a bee in one minute.

Knapweed flower with pollinating hover fly.

Compare the numbers and types of insect which visit two different types of composite flower such as the cow parsnip (Umbelliferae) and the Dandelion (Compositae) over a whole season.

For flower structure and pollination mechanisms see Maud Jepson *Biological Drawings* and M. Proctor and P. Yeo *The Pollination of Flowers*.

98   OE   Garden Birds

Choose one or more of the following species: house sparrow, hedge sparrow, robin, blackbird, song thrush, starling. Compose a hundred questions you would like to answer about their habits from your own observations. Later you can check the answers in books, magazines or encyclopaedias but the chances are that you will make discoveries that are not to be found in any of these places. A pair of binoculars is needed, regular feeding of the birds in winter is recommended. A portable tape recorder is useful.

Think out your questions under subject headlines such as plumage, feeding, movement, territory, song, nest, eggs, young, and so on. Examples might be:

1   Does the cock differ from the hen in plumage, if so in what ways? (A reference book might have to be the starting point here.)
2   At what times of day are the birds found feeding,   a) if food is put down for them,   b) if no food is put down?

Make drawings and/or take photographs to illustrate your project.

99   AO(E)   A study of web-making spiders in the house and garden

This is a suitable study for those without unreasonable fears as spiders are common through most of the year and are varied and interesting in their habits.

There is a good account in Michael Chinery *The Natural History of the Garden*, while more detailed accounts of the species are to be found in W. S. Bristowe the *World of Spiders*.

Look for webs strung along fences and across doors and windows, draped over walls and bushes or carpeting the grass. Observe funnel shaped webs in crevices in walls and tree trunks and triangular platforms of cobwebs in every neglected corner of the shed, greenhouse or home.

The diadem or garden spider *Araneus diadematus* is one which builds a beautiful orb web, the construction of which is described in both the books mentioned. Look for this in late summer and autumn evenings. Try to describe the stages in the construction of the web by sketches, photographs and writing. Check this with descriptions in the books, then look again to observe the points you have missed. To see the spider in the centre of the web, go out at night with a torch. During the daytime she is usually sitting in a hidden place holding a connecting thread. Watch for her actions when a flying insect blunders into the net or when you imitate such a happening by vibrating the web. Watch the biting and wrapping of the prey in silk and what happens if dangerous or vigorous insects land in the web. Try the effect of introducing insects yourself, such as wasps or locust hoppers. Is the prey sucked immediately for food or is it carried away for consumption later?

Orb webs found in late summer and autumn in the lower parts of the hedgerow and in the herbaceous border, with no central platform and slung at an angle of 30–70° from the horizontal, are probably those of *Meta segmentata*. Webs found on window frames and garden sheds may be those of *Zygiella X notata*. The young make orb webs, but the adults miss the spiral threads from two sectors and the webs are often asymmetrical with many more sticky threads below the hub than above. There is also a tubular retreat.

Look for the white lacy webs of the three *Amaurobius* species on old walls and in sheds. There is a hole in the centre leading into a crevice, which is the spider's retreat. Pull the web out of the crevice to reveal what the spider has been eating and to see how easily the lacy threads stick to your fingers. Sort out the remains of flies, earwigs, woodlice etc., and cast spider skins. Make these into permanent slides as described in the Appendix. If you find small pink spiders in the web these are *Oonops pulcher*, an iniquiline, which is an associated and harmless animal. *Segestria senoculata* is also to be found on walls, forming a silken tube in a cavity between bricks and stones. There are radiating trip wires.

There are three common domestic species of House Spider, *Tegenaria domestica*, *T. atrica* and *T. parietina*. These are the long-legged spiders that form hammocks or cobwebs in corners of rooms and sheds and cause alarm in the house. The first has the thickest and whitest sheet web of the three and reaches a body length of ten mm. The second makes her way in autumn from the outside by way of overflow pipes from bathrooms and other inlets. The third is as large as the second, but the legs are much longer and more furry and it is found mostly in southern England. Instead of fleeing in horror from any of these, persuade one to run into a small glass specimen tube. When stoppered the spider can be examined closely and identified. It can be transferred to a larger jar with a nylon or muslin top and kept for a few days without food and released, or alternatively kept for longer if fed with flies. (There are sometimes surplus *Drosophila* flies from genetic experiments in the biology laboratory which can be used.) Do not keep two spiders in one container and do not place the container in the sun. A tank with damp earth or foliage kept in the shade is best for outdoor species.

Other web-building spiders include *Theridion sisyphium* found from June onwards in the herbaceous border, with a web like a three dimensional trellis. The shrubs and hedgerows may be adorned in autumn with the slightly domed sheets of *Linyphia triangularis*; the spider hangs upside down from the lower surface. This species can easily be persuaded to construct a web if kept in an insect cage (see p. 84) with a forked twig in a jar.

To make a collection to illustrate your project, spray webs with white paint from an aerosol to make them more visible and collect them on black cardboard. Cover with clear adhesive material such as Fablon. Label carefully with species and place found. Such a collection should be supplemented with photographs and sketches of the whole web and stages in its construction.

If you wish to preserve one spider of each species to check identification and for further study, kill them by dabbing a little ethyl acetate on the inside of the cork of a specimen tube in which you have caught them. Examine and preserve in 70% ethanol.

Live spiders are very suitable for use in choice chamber experiments to test preferences for light, dark, damp, dry, cold, hot, and so on. For the large species you will need to design a different version of the usual petri dish type of choice chamber.

100   AOE   Habits of garden molluscs

First survey the garden in summer with the help of a good reference book like Edward Step *Shell Life*, Michael Chinery the *Natural History of the Garden* or the *Oxford Book of Invertebrates*. Decide which species of snails and slugs are present and which of these are abundant. Best places to look are amongst clumps of plants in the rockery and herbaceous borders, on leaves of cabbages and other vegetables, amongst long grass, under stones and pieces of wood, in piles of flowerpots or underneath pots and garden containers. Choose either snails or slugs, or a common species of either group to study in detail over a period of a year if possible. A shorter project could be carried out over the period of the summer holidays.

If you choose for example the garden snail *Helix aspersa*, find a favourite roosting place in which they spend most of the day and from which they emerge on their food finding expeditions in the evening. Mark a number of snails with a blob of waterproof paint. Remove them to the other end of the garden and examine the roosting place again next morning. You will probably find your snails have returned. Edward Step describes in *Shell Life* how he simply pencilled his initials on one such snail and hurled it as far as he could. Next morning it had returned to its original roost although it had to cross a very broad road and climb a low wall! Throwing snails is not to be recommended however as they may break in the process and if they land in your neighbour's garden this may make you very unpopular. Carry out controlled experiments and using the scientific method, test from how far snails will return to their roosts.

Collect as many snails as you can and place them in a large tank or several small tanks in a shady part of the garden. Old kitchen sinks, aquaria tanks (even old cracked ones which will no longer hold water) and smaller plastic tanks are suitable. Put some damp soil at the bottom, some stones and pieces of wood for cover and a few plants if you like. Cover with a snail-proof lid of wood or stone or perforated zinc. Place about six snails in a small tank or twelve in a larger one.

Test food preferences using a) vegetable leaves, b) weed leaves, c) leaves of herbaceous plants. On this basis and after consultation with the gardener, decide how to feed your snails over the season. Find out how much the snails eat per day, how they eat and at what time of the day or night they eat. Remove all stale leaves before they decay.

Observe the formation of the epiphragm, a layer of hardened mucus which closes the opening of the shell preventing evaporation in dry weather or during the winter. Record the date at which a much thicker epiphragm forms to keep out the cold during the winter hibernation. Record the period of hibernation during which the snails do not feed. Try the effect of keeping some in warmer conditions in a shed or the school laboratory. Look out for eggs in the spring, count them and see how many hatch and how many grow throughout the season. It may be possible to construct growth curves for the young snails.

If living in areas where the related Roman snail *Helix pomatia* is found, such as the North Downs in Kent, it is possible to compare feeding, hibernation, breeding and so on with the common snail for an A level project.

(*101*) *A(OE)   The ecology of a bird bath with special reference to the rotifer,* Philodina roseola

If you have an old established concrete or stone bird bath in the garden you can be off to a good start with this project, either on a short or long term exploration. If you have no bird bath, try sampling gutters which are easily accessible. You also need access to a good microscope.

Start with a bird bath that has not been recently cleaned or refilled. Record any seeds or feathers present which will help you to know which birds have visited the bath and what they have been eating. Put a drop of water from the bath and some of the pinkish sediment scraped from the sides on to a microscope slide. Carefully add a coverslip without introducing air bubbles. Examine under low and then high power. You are likely to see unicellular algae, which include the red *Sphaerella lacustris*, filamentous algae, bacteria, amoebae and other protozoa and the spectacular rotifer, *Philodina roseola*. There is a coloured picture of this in the *Observer's Book of Pond Life*. This wheel animalicule exhibits the phenomenon of anabiosis. Under dry conditions it contracts its head and foot until the body is approximately spherical. It secretes mucus around itself which forms a hard impervious surface when dry. In this form it is carried on the feet of mammals and birds or blown by the wind and can survive for years. As soon as it falls into water the mucus coat softens and it becomes active again.

Make accurate scale sketches of all the organisms found and notes as to colours, movement and feeding.

Repeat this procedure at different seasons of the year, keeping a note of maximum and minimum temperatures. Also repeat at times when the bath has been cleaned and refilled and when it has dried out. In the latter case, *Philodina* may be found in the encysted form. Watch its revival in a drop of water on a microscope slide over a period of hours. Conversely, watch the encystment of an active animal by withdrawing water from the coverslip by means of blotting or filter paper. Make sketches at intervals to record both processes.

To make this project quantitative over a period of a year, when the bath is about half full use a stiff brush to detach organisms from the sides and base of the bird bath. Swish the water round but do not swish it over and take a sample with an old tablespoon and tip this into a jam jar or similar container. Subsample this by stirring with a graduated pipette, transferring a measured amount of this on to a clean glass slide. The amount taken should be just enough to make a thin film below a standard sized coverslip. Alternately it might be possible to adapt a haematocytometer slide, used for counting blood corpuscles for the purpose. Choose one or more types of organism which is abundant and which you can identify. If using an ordinary slide, count the number of these organisms in three successive fields of view. Whether you choose a low or high power field will depend on the abundance and size of the chosen organisms. Repeat this standard procedure at different times of year and relate the results from the mean readings to time of year and any other factors that you decide are relevant.

102   OE   The ivy Hedera helix *(climbing, variation, pollination and dispersal)*

This common plant can climb to 50 m or creep along the ground forming carpets, flowering only in sun at the top of whatever it is climbing on.

Study this plant in your own garden, in the school grounds, on the walls of houses and in parks. Compare its growth under these different conditions. Look for the yellow green flowers in late autumn and the purplish black berries in early spring.

Choose one plant, preferably one that flowers and fruits, to study in detail. Can you distinguish between the lobed and pointed leaves and the different positions they occupy on the plant? See how many variations you can find in leaf colour, vein colour, leaf shape etc. Observe and draw the adventitious roots which are used for support not nourishment, as all the minerals and water from the plant are obtained from the ground. Look for the flowers in late September. The greenish yellow disc in the centre of each flower secretes nectar which attracts flies and wasps. Collect and identify these if possible. Watch for the ripening fruits in spring and record which birds eat them. Examine the small hard seeds and find out whether these are swallowed by birds and in what manner they are dispersed.

In spring, measure the rate of growth of the new shoots, leaves and adventitious roots and relate this to weather conditions, especially temperature.

Observe the leaf mosaic pattern made by ivy on a wall. Record by sketches or photographs. How does it come about, and what is its advantage to the plant.

103 *(A)OE* *Annual and perennial weeds in the garden*

By definition weeds are simply plants growing where the gardener or farmer does not want them. In the hedgerow the same plants may be considered as wild flowers, admired for their beauty. With the disappearance of many of our wild flowers, some gardeners are learning to live with their weeds or to keep a part of the garden where they may grow and where the butterflies associated with them (such as the tortoiseshell and the peacock with the stinging nettle) are encouraged.

Compare the habits of growth and the success of the annual and perennial weeds which happen to be present. Annuals are likely to include groundsel, shepherds purse, chickweed, poppy, scarlet pimpernel, speedwell. Perennials include dandelion, daisy, sow thistle, creeping thistle, creeping buttercup, stinging nettle, dock and plantains. Identify any of these weeds or others present in the garden and classify them as annuals or perennials. Uproot one plant of each species when in flower and draw to show the features which adapt the plant for its success as a weed. In annuals this might include numerous wind dispersed fruits such as those of the groundsel, and in perennials the long root of the dandelion, so difficult to eradicate.

Clear all weeds from a certain part of the garden, recording the species and the relative abundance of those removed. This part will be popular with the gardener of the house, providing you really know your weeds. For the next part you will have to plead the interests of science and the importance of your project. Leave the cleared plot for as long as you are allowed (several months if possible), and record each species as it returns. Thus if the plot is cleared in June, you would record: first week of July two groundsel, one creeping buttercup; second week of July ten groundsel, three creeping buttercup, one dandelion, and so on.

The project could develop in various ways, according to the time available and the interests of the students, for example the time taken to complete the life cycle of chosen annuals and perennials, a comparison of the methods of propagation of the two groups, ease of eradication and the effectiveness of chemical control of the two groups or different species within the group.

Further Reading

1   Rolls, M. J. and Morrison, R. B. *Science in the Garden* Blandford 1966
2   Bristowe, W. S. *The World of Spiders* Collins New Naturalist 1971
3   Burton, J. *The Naturalist in London* David and Charles 1974
4   Chinery, M. *The Natural History of the Garden* Collins 1977
5   Clarke, L. *Botany as an Experimental Science in Laboratory and Garden* O.U.P. 1935
6   Mitchell, A. F. *Three Forest Climbers* (Ivy, Old Man's Beard and Honeysuckle) Forestry Commission
7   Mourier, H. Winding, O. and Sunesen, E. *Collin's Guide to Wildlife in the House and Home* Collins 1977
8   Needham, Frost and Tothill *Leaf Mining Insects* 1928
9   Proctor, M. and Yeo, P. *The Pollination of Flowers* Collins New Naturalist 1973
10   Nuffield Teachers' Guide *Introducing Living Things* Text 1 Longman 1966 (See note on p. 43)
11   Rowett, H. G. O. *The Invertebrates* (Dissection Guides) Murray
12   Russel, Sir William *The World of the Soil* Collins New Naturalist No. 35 1961
13   Salisbury, Sir Edward *Weeds and Aliens* Collins New Naturalist 1961

14 Soper, Tony *The Bird Table Book* David and Charles 1965
15 Soper, Tony *Everyday Birds* David and Charles 1976
16 Stainton, H. T. *The Natural History of the Tineina* 1855–73 (13 vols)

See also titles on Further Reading Lists: Chapter 1 Nos. 4, 6, 16 and 17; Chapter 2 Nos. 3, 10 and 12; Chapter 4 Nos. 1, 5, 6, 7, 17, 19, 26, 30 and 31; Chapter 5 Nos. 3, 9, 10, 11, 12, 13 and 14

Appendix

Detailed methods of work

1 Introduction

Ecological methods which are fully described in biology textbooks or common reference books are not described here. Some of these books are included in this reference section even if they have been listed for earlier chapters.

The catalogues of biological suppliers such as Philip Harris and Griffin contain much sensible and up to date advice in their fieldwork sections, especially about quantitative methods. Certain apparatus and equipment listed will be found very useful for some of these projects. The recently developed equipment for monitoring environmental parameters, such as oxygen, sound, light, conductivity and pH are highly desirable items if funds permit. In many cases simple apparatus can be devised and made by pupils. Examiners or project assessors will naturally take such enterprise into account when judging projects.

The results from some projects may require statistical analysis especially where large samples are involved and where the work is of A level standard. Some books giving advice on this aspect of the work are included in the references. They should be consulted *before* planning the details of projects involving quantitative work and random sampling.

The methods of work described below are those which have been found particularly helpful, which are relevant here and which may not be described in readily available books.

2 Collecting and preserving plants

The Conservation of Wild Creatures and Wild Plant Act, 1975, makes it an offence to pick twenty-one endangered plant species, none of which students are likely to come across in the projects suggested here. It is also an offence to uproot any wild plant without the permission of the owner of the land. Only collect plants when it is essential to your project, using the advice in Project 3. Most plants are more easily identified when growing or fresh, but sometimes it may be necessary to preserve plants until a positive identification can be made.

A suitable plant press can be bought from biological suppliers or can easily be made (Fig. 1). The overall dimensions are 350 × 450 mm. Make the press from two pieces, each composed of plywood strips firmly screwed and glued together as shown in the

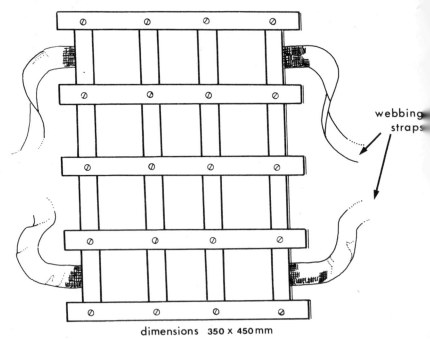

webbing
straps

dimensions 350 x 450 mm

Fig. 1   Plant Press

diagram. Use strong webbing straps with buckles to tighten the press. Fill the press with large double sheets of newspaper folded in halves and place a strong piece of card between the newspaper and each of the two frames.

As you pick each plant, place it carefully between one of the folded newspaper pieces arranging it so that it will press to the best advantage showing details of leaves, stem, buds, flowers and fruits. If the plant is tall cut it into convenient sized portions. Slice large fruits and stems vertically. Tie a label on to the plant securely and record all relevant details in your notebook. These should give the exact location, height of plant, habit (whether climbing, prostrate or erect), colour and scent of flower (lost during pressing), or leaves and any other relevant information. When all your collecting is done, tighten the straps as much as possible.

On return to base, place the press in a warm dry place but not by direct heat. After a few days, change the newspaper. Repeat this

Plant press and mounted plants. (Lake Nakuru, Kenya)

until the plants are completely dry and flattened. This will take a week or longer according to weather conditions. In this form the plants can be identified by using a Flora or by checking with an expert. Do not go entirely by pictures for identification but use the key and read the description carefully. This will prevent the identification of a common hedgerow plant as one that only occurs at the top of Ben Nevis, to use an extreme example!

If you wish to mount your specimens, use pieces of thin card or thick paper of the same size as your plant press. Use a good PVA adhesive. Place a few drops of the glue on the under surface of each leaf, fruit or wide stem and arrange carefully on the sheet. Apply pressure with a sand bag or suitable weight placed on top of the specimen, or pile of specimens. Stick on a large label at the lower right hand corner giving full details.

Place each finished sheet in a polythene bag or folder containing a few crystals of paradichlorbenzene (obtainable from large chemists). This will protect against the depredation of insects. Thorough drying protects from fungal attack.

3 Collecting and preserving small invertebrates

Wherever possible study and identify these animals and return them to the exact habitat as soon as possible. Where identification is not possible from the live animals it is better to kill a few selected animals by the methods given than to collect extensively and neglect, forget, or throw away the whole collection.

If a killing bottle is needed use the type with chopped cherry laurel leaves which are then crushed with a hammer and put in a glass tube or jar beneath paper or cloth, the insects being placed on top and the lid closed. The method is slow but effective. It keeps insects relaxed indefinitely but is not suitable for large beetles.

Most small invertebrates, including insects may be killed quickly by dropping them into undiluted ethanol. Beetles should be placed in a small glass tube containing cotton wool soaked in ethyl acetate and left for about an hour. Small crustaceans are best killed in dilute formalin. Land and water snails, fly larvae and earthworms may be killed by a short immersion in boiling water, long enough for them to stop moving. After cooling, remove the soft parts of snails with a pin. Place the shells in small tubes with cotton wool plugs at each end so that the shells do not move easily, enclosing a label with details of habitat and identification. Mount beetles on card with a drop of glue. Preserve other small arthropods in small stoppered tubes of 70% ethanol kept inverted in a large open mouthed bottle of 70% ethanol. This prevents evaporation of the contents.

Very small arthropods such as mites and springtails may be killed and preserved by mounting in Hoyer's medium as described in the next section. For further details consult Oldroyd *Collecting and Preserving Insects*, the *British Museum Instructions for Collectors* and Ford *Studying Insects*. The latter book might provide many other ideas for projects.

4 Preparation and use of Hoyer's Medium

Use this medium for mounting small animals such as insect larvae, nematodes, mites, springtails and small beetles from soil, leaf litter, moss or fungi. Slides made in this way are semi-permanent, remaining unchanged for five years or more if carefully prepared.

The teacher or laboratory assistant should prepare the medium. Use it under supervision in the school laboratory as it contains poisonous chloral hydrate.

Dissolve 30 g of clear crystals of gum arabic in 50 g water. Add 200 g chloral hydrate, dissolve this thoroughly, then add 20 g glycerine. Use from dropping bottle.

Use ordinary glass microscope slides for very small flat specimens such as springtails and nematodes. Use cavity slides for hard bodied mites and beetles or larger larvae. Place enough Hoyer's Medium to fill the cavity and the space below the coverslip, being sure that it is free from air bubbles. Place the animal in the centre and lower the coverslip with a mounted needle so that the medium spreads slowly and evenly. Label the slide fully.

Mount moss leaves directly in the medium. Any small organisms attached to the leaf will be killed and all but the most delicate ones will be preserved. Chloroplasts are not preserved by this method.

The Hoyer's medium not only preserves the plant or animal tissues but after a few days it clears them beautifully so that it is possible to see what springtails, softbodied mites and larvae have been feeding on. Pollen grains, fungal spores and other structures can be identified in the gut.

Store the slides flat in a covered box for a month until the medium sets completely, then ring the coverslip with colourless nail varnish. The slides may then be stored edgeways in a slide box.

Drawings can be made from the slides at a convenient time and the collection used for comparing the various organisms found. Animals and plants preserved in this way can be recorded even more permanently by means of photo-micrography. A reference book is given for those interested.

5 Random sampling

In some projects the choice of biological material for study is subjective. For instance, a rock pool of suitable size, easily accessible and likely to yield interesting results as in Project 18. In other projects the 'choice' must be made as objective as possible, the student eliminating his own preferences. It is important to understand the principles involved in random sampling and the methods by which it may be carried out.

The purpose in each case, is to make sure that each plant or animal or quadrat area has an equal chance of being chosen for study each time that sampling is carried out. This is necessary if, for example, a certain number of leaves are chosen from a tree in order

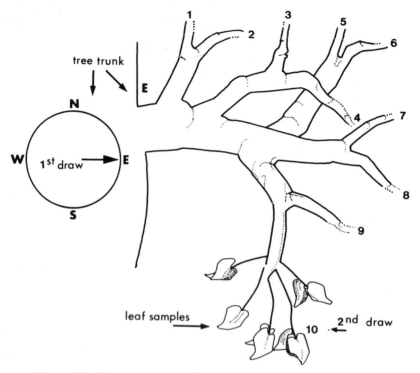

Fig. 2   Random sampling of leaves. (Projects 53 and 66)

to estimate total leaf weight as in Project 53 (Fig. 2). If quadrats inside and outside an exclosure are selected as in Project 80 this must also be done randomly.

It is not proposed to give here all the possible, detailed methods of random sampling to fit every project. The person who assesses your project for examination or other purposes will be interested in how you solve such a problem and the way in which you justify the method chosen. A few examples and pointers are given here to show the general idea. Students may think of better methods.

In Project 53 the simplest random method of selecting the leaves or fruits might be to draw lots by putting them in a paper bag and pulling out the first ten that come to hand, without feeling about or deliberately taking the largest or smallest.

When selecting a random quadrat inside the exclosure in Project

80, give each metre a letter, A, B, C, D, E along one side and numbers 1, 2, 3, 4, 5 along an adjacent side. Select a letter from one bag and the number from another and so locate the quadrat position, as say B4, and measure out along the ground accordingly. For choosing the quadrat position outside the exclosure, use three bags, the extra ones having numbers 1–10 or more, indicating the number of metres or half metres away from the exclosure. Alternatively a table of random numbers, such as may be found in statistical tables, may be used in any appropriate way that does not involve subjective choice. For example the first part of the first line of the random number table runs:

1698 4865 6464 2231 3163 7736 0822 5872 6979 2339

Selecting the first trio of numbers, 223, which fits the case, locate the quadrat as shown in Fig. 3. The next usable trio might be 316 and then 233 and so on.

Project 77 requires the selection of ten nettle plants from the same patch each week. Put down a measuring tape at the edge of the

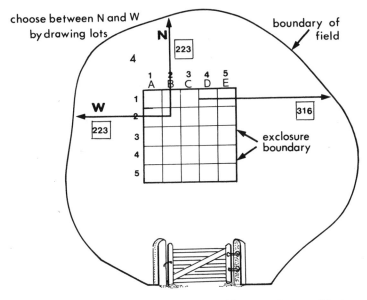

Fig. 3 Random choice of a quadrat outside an exclosure. (Projects 50 and 80)

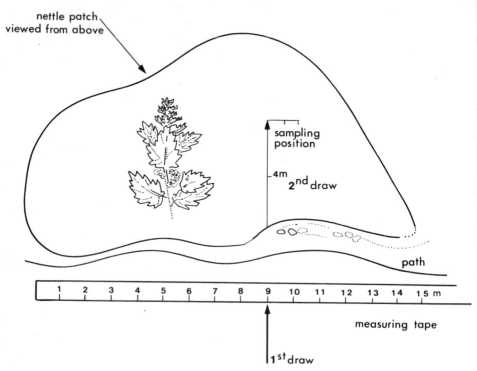

Fig. 4   Random sampling of nettles. (Project 77)

patch and draw lots for a number of metres, say from 1–10, according to the size of the patch. Draw lots again for a number of metres inside the patch. With the help of an assistant, measure the height of each nettle plant which touches a decimetre mark when the metre rule is held with the right hand extended to the side of the body (Fig. 4). Wear suitably protective clothing!

Collecting ten berries at random from several bramble bushes requires more ingenuity. One way is to locate a branch, drawing lots to give height from the ground and distances from front to back, or left to right. Pick all the ripe berries from the branch (providing it is reachable!), put these into a bag and pick out ten without looking (Fig. 5).

To select water lily leaves at random as in Project 28a, choose a transect across the river where there are most water lilies. Always

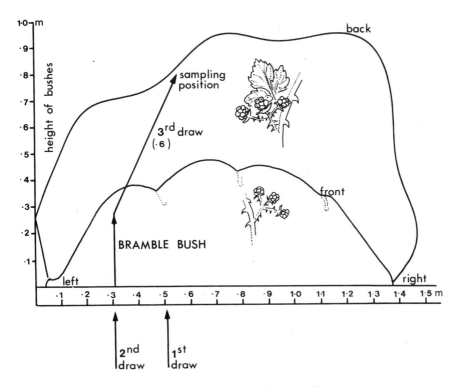

Fig. 5   Random sampling of blackberries. (Project 76)

start from a fixed point on the bank and choose the leaf nearest to a
decimetre number on a marked line or measuring tape. Choose the
numbers by drawing lots or from a random number table (Fig. 6).

6 Quadrats and transects
These provide useful quantitative methods for comparing flora
and/or fauna from different areas and are frequently mentioned in
the projects. They are described in detail in John Sankey's excellent
Guide to Field Biology as well as in other ecology textbooks and field
guides.

  Two forms of quadrat frame not usually mentioned have been
found especially useful. First is the simple half metre square
wooden frame divided into decimetre squares by tightly drawn and

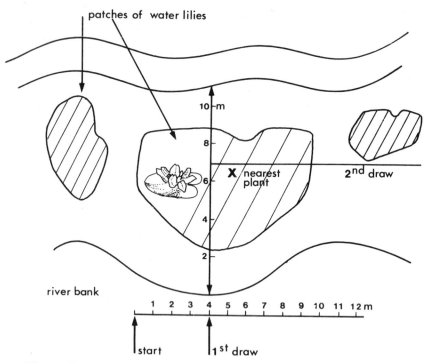

Fig. 6   Random sampling of water lilies (viewed from above). (Project 28a)

Half metre quadrat frame for studying weeds in the lawn.

Close-up of half metre quadrat frame (1 decimetre²).

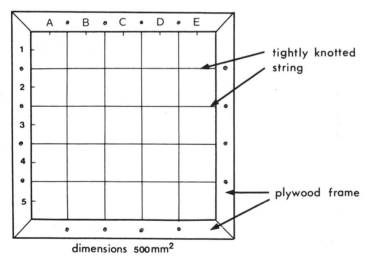

Fig. 7 Quadrat frame.

knotted strings. This is easily made, light and versatile (Fig. 7). It is suitable for most of the projects in this book which involve quadrats. Standard wooden or plastic expanding trellis is useful for

Use of trellis as quadrat frame.

Detail of trellis used as quadrat frame.

both quadrats and belt transects. Its advantages are that it is easily portable, can be bought in several convenient sizes and the small squares of which it is composed are approximately decimetre squares. They can be used on flat, sloping or irregular ground, or on water. The wooden ones tend to be rather fragile in use if not carefully stored and handled. The cost is fairly high but compares favourably with frames bought from biological suppliers. Point quadrat frames, if required, can be made by boring small holes at decimetre intervals in a metre rule and inserting long knitting needles but in many cases as in Project 77 a metre rule on its own is adequate for selecting plants for measurement or recording within a patch or quadrat.

For a line transect you need a measuring tape marked in metres and decimetres, a spirit level with sights and measuring poles marked in feet or decimetres. This enables two people working together to obtain the correct shape of the profile. For a belt transect use a piece of expanding trellis or two parallel measuring lines plus a quadrat frame. Record quadrat and transect information in a notebook or on squared paper using a different symbol and key for each species, or shading to represent percentage cover of vegetation in each decimetre square.

7 Dissolved oxygen estimation in aquatic habitats

1 Constant recording oxygen probe and recorder
The most informative method of measuring the percentage of dissolved oxygen in freshwater habitats is by a constantly recording oxygen probe. Students can then appreciate the daily fluctuations and the effects of pollution over a long period of time and record it accurately with the advantage that it is automatically corrected for temperature. The cost will be sufficient deterrent for most schools, but there may be ways of securing grants as the present author was able to do for such equipment from the Royal Society's scheme for scientific research in schools. The problems involved in setting up such apparatus in the field are considerable, but if they can be solved the results obtained are well worth it (Fig. 8).

2 Portable probe and meter
An easy and accurate way of making spot checks of dissolved oxygen in freshwater and marine habitats is by means of one of the portable probes and meters now on the market and available through biological suppliers. These also have the advantage of

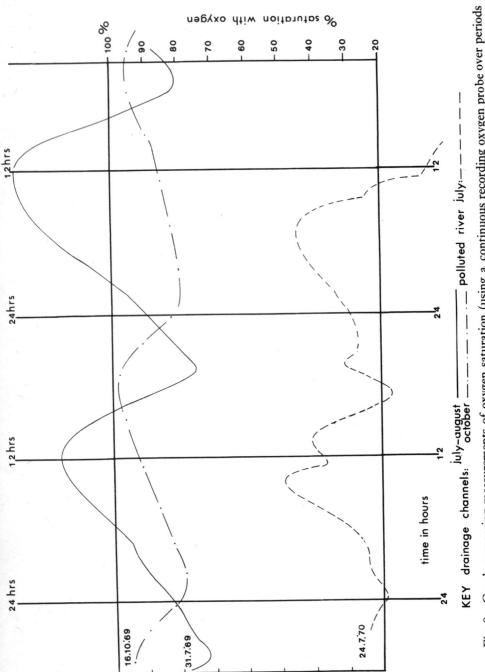

Fig. 8   Graph comparing measurements of oxygen saturation (using a continuous recording oxygen probe over periods of 48 hours) in recently dredged drainage channels and a polluted river on the Pevensey Levels, East Sussex.

KEY drainage channels: july–august ————
                         october ————·————   polluted river july:— — — — —

being adaptable to measure other parameters if bought as part of the environmental monitoring equipment described in the Introduction. Again the cost is fairly high.

3 *Winkler's technique*
This chemical technique is well described in several books. It is somewhat time consuming for most school projects and must be carried out immediately after the field expedition.

4 *Modified Winkler's technique*
This technique described by B. F. Gill in Vol. 58, No. 204, March 1977 of the *School Science Review* is a useful simplification that can be used in the field for O and A level projects. (Fig. 9.)

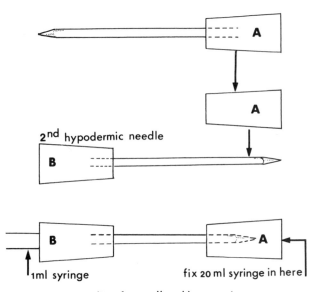

2nd hypodermic needle

1ml syringe   fix 20 ml syringe in here

coupling for small and large syringe

Fig. 9 Coupling device for modified Winkler's technique for estimating dissolved oxygen in fresh or salt water under field conditions.
1. Remove the plastic fitting A from a hypodermic needle by gripping the needle in a vice and using pliers.
2. Push fitting A onto a second needle B.

REAGENTS

A   40% w/v manganese (II) chloride-4-water solution.
B   A solution of 80 g sodium hydroxide and 2.5 g potassium iodide in 250 cm³ water.
C   Phosphoric (V) acid—about a 50% solution.
D   0.025 M(M/40) sodium thiosulphate (VI) solution.
E   0.25% w/v starch in saturated sodium chloride solution.

PROCEDURE

1   Take up a 14 cm³ water sample in a 20 cm³ plastic syringe (a needle is not required). To avoid getting air bubbles into the sample proceed as follows. Take up about 20 cm³ *slowly*. Invert the syringe and tap it to bring any air bubbles to the nozzle. Expel any air and excess water until *exactly* 14 cm³ remain.
2   Take up a small amount of reagent A. The amount is not critical but a nozzle full is convenient. Wipe the nozzle.
3   Take up a similar amount of reagent B. A precipitate of manganese (II) hydroxide will form as the syringe is gently turned to mix the contents. Wipe the nozzle.
4   Wait for at least a minute so that the precipitate of manganese (II) hydroxide can take up oxygen.
5   Take up reagent C in the same way as A and B. The precipitate will dissolve, liberating iodine. Draw a few cm³ of air into the syringe.
6   Fill a 1 cm³ plastic syringe (no needle) with reagent D. Invert, remove air bubbles and expel to the 1 cm³ mark. Fit the small syringe into the large one with the coupling device (Fig. 9). The 20 cm³ syringe should be uppermost.
7   Inject thiosulphate (VI) solution until the iodine is a pale straw colour.
8   Detach the 1 cm³ syringe and take up a few drops of reagent E into the 20 cm³ syringe. A blue-black coloration will develop.
9   Attach the 1 cm³ syringe and continue to inject thiosulphate until the blue-black coloration disappears; this is the end point.
10 Note the amount of thiosulphate (VI) *which has been injected*. The oxygen concentration of the water sample is the amount of thiosulphate (VI) used, multiplied by 10; that is, if 0.63 cm³ of thiosulphate were required then the oxygen concentration of the water sample was 6.3 cm³/dm⁻³ water.

Pack the reagents carefully and label them clearly. Practise the
technique several times before using it in the field.

8 Recording relative humidity

Rough comparisons may be made by timing the rate at which dry
blue cobalt chloride paper (carried in a tube containing a little silica
gel or calcium chloride) turns pink. A standard piece of damp pink
cobalt chloride paper covered firmly with adhesive tape is useful for
comparison.

A wet and dry thermometer provides a more accurate method.
This consists of two thermometers mounted close together, the wet
bulb being covered by a piece of saturated muslin dipping into a
water reservoir. As evaporation causes cooling, the greater the
difference between temperature readings after no further change is
observed, the less the humidity. The difference can be quoted as an
arbitrary number, or converted to relative humidity by reference to
Meteorological Tables.

Further Reading

1   British Ecological Society Symposium No. 7 *The Teaching of Ecology*
    Blackwell Scientific Publication 1966
2   British Museum (Natural History) *Insects* (Instructions for Collectors
    No. 4A) 1974
3   British Museum (Natural History) *Invertebrate animals other than
    insects* (Instructions for Collectors No. 9A) 1954
4   Elliott, J. M. *Some methods for the statistical analysis of samples of
    benthic invertebrates* Freshwater Biological Association Scientific
    Publication No. 25 1971
5   Ford, R. L. E. *Studying Insects* Warne 1973
6   Kodak Publication *Photography through the microscope* No. P-2 1974
7   Oldroyd, H. *Collecting, Preserving and Studying Insects* Hutchinson
    1970
8   Parker, R. E. *Introductory Statistics for Biology* Studies in Biology
    No. 43 Edward Arnold 1973
9   Sankey, J. *A Guide to Field Biology* Longman 1958
10 Schwoerbel, J. *Methods of Hydrobiology* (Freshwater Biology)
    Pergamon Press 1970

Index

F. excelsior 75
Freshwater Biological Association 24
  key 31
frog-hoppers 98
fronds 81
fruits 61, 97, 98, 104, 113–115, 117, 124,
  127, 134, 141
Fucus serratus 11, 14
Fucus spp. 7, 12
F. vesiculosus 14
Funaria hygrometrica 90
Fungi 61–62, 63–72, 90, 91, 92, 109–110,
  111, 147, 148
fungus foray 63, 68

Galium saxatile 55
gall gnat 102
galls 102, 111
Ganodermus applanatum 64, 67
garden snail 138–139
garden spider 136
gastropods 36, 40
gastrotrichs 75
genetic factors 112
Geotrupes 109
germination 61, 81, 104
Gibbula spp. 13
gill prints 64
gills of fungi 64
girdle scars 59, 62
glycerine 149
goosegrass 116
gorse 52, 69, 114, 133
gradient 49
grasshoppers 107–108
grassland 54, 107
great pond snail 33
green bottles 99
Griffin apparatus 145
groundsel 113, 141, 142
grouse moor 54, 55
growth 33, 59–62
  curves 102, 139
Guides 1
gum arabic 149
guttering 92
gutters 139

haematocytometer 140
hardness of water 27
Harris apparatus 145
hawthorn 114, 128
hazel 83, 114, 128
heath bedstraw 55
heath, cross-leaved 55

fine-leaved 55
  sedge 55
heather 52, 55, 56
  beetle 55
heather-moor 54
heaths 46
Hedera helix 140–141
hedgerows 80, 97, 115–117, 136, 147
hedge sparrow 135
Helix aspersa 138–139
  pomatia 139
hen 135
herbicides 40
herbivores 61, 68, 106
hibernation 88, 139
hide, bird 7
Himanthalia elongata 11
hive bees 99, 134
holiday projects 2, 4, 9, 43, 119
holly 62, 114
  leaf miner 129
honeysuckle 129
hornbeam 113
horsechestnut 59, 114
house sparrow 135
  spider 136
hover fly 98, 134
Hoyer's medium 67, 68, 73, 75, 104,
  148–149
humidity 9, 19, 80, 91, 102, 116, 161
humus 34, 81, 100, 124
hydroids 13
hydrometer 17
hydrosere 39
Hylocomium screberi 55
hyperparasites 111
Hypnum cupressiforme 55
hypodermic needle 159

incident energy 61
incubation 110
incubator 60
indicator organisms 33
indumentum 101
inflorescence 97, 101, 113
iniquilines 111, 136
insecticide 110
insects 18, 31, 33, 34, 40, 53, 55, 61, 62,
  67–72, 82–86, 102, 106–111, 115, 117,
  128–131, 134–135, 136, 147, 148
internodes 52, 75–77
intestines 53
invertebrates 55, 82, 91, 130, 148
iodine 160
Isle of Man 4